8th Grade Mathematics

This Test Review Booklet was designed for Grade 8 Mathematics Assessment Test. It provides examples of the format and types of questions that may be on the actual test as administered by the State Education Department. We have separated our review tests into two sections:

The actual test has three books, administered over three days.

Part 1:   28 multiple choice questions

Part 2:   27 multiple choice questions

Part 3:   6 short response questions
4 extended response questions

For a complete description of restrictions see the NY State Education website: www.nysed.gov

Special Thanks To:

Luke Masouras - Examgen Inc.

Syracuse, NY • www.EXAMgen.com

For providing technical guidance and development of the test questions

© 2016, Topical Review Book Company, Inc. All rights reserved.
P. O. Box 328
Onsted, MI. 49265-0328

Grade 8 Mathematics Reference Sheet

CONVERSIONS

1 inch = 2.54 centimeters
1 meter = 39.37 inches
1 mile = 5,280 feet
1 mile = 1,760 yards
1 mile = 1.609 kilometers

1 kilometer = 0.62 mile
1 pound = 16 ounces
1 pound = 0.454 kilogram
1 kilogram = 2.2 pounds
1 ton = 2,000 pounds

1 cup = 8 fluid ounces
1 pint = 2 cups
1 quart = 2 pints
1 gallon = 4 quarts
1 gallon = 3.785 liters
1 liter = 0.264 gallon
1 liter = 1,000 cubic centimeters

FORMULAS

Triangle	$A = \frac{1}{2}bh$
Parallelogram	$A = bh$
Circle	$A = \pi r^2$
Circle	$C = \pi d$ or $C = 2\pi r$
General Prisms	$V = Bh$
Cylinder	$V = \pi r^2 h$
Sphere	$V = \frac{4}{3}\pi r^3$
Cone	$V = \frac{1}{3}\pi r^2 h$
Pythagorean Theorem	$a^2 + b^2 = c^2$

8th Grade Mathematics
TABLE OF CONTENTS

Published by

TOPICAL REVIEW BOOK COMPANY

P. O. Box 328

Onsted, MI. 49265-0328

EXAM   PAGE

1 What is the value of $\sqrt{1}$?

A 0

B 1

C $\frac{1}{2}$

D −1

1 _____

2 What is 48,200 written in scientific notation?

A 4.82×10^3

B 4.82×10^4

C 4.82×10^2

D 4.82×10^5

2 _____

3 What is the sum of 6×10^3 and 3×10^2?

A 6.3×10^3

B 9×10^5

C 9×10^6

D 18×10^5

3 _____

4 Which of the following equations has an infinite number of solutions?

A $3(2x + 1) = 2(1 + 3x) + 1$

B $3(2x + 1) = 2(1 + 3x) − 1$

C $3(2x − 1) = 2(1 + 3x) − 1$

D $3(2x + 1) = 2(3 + 2x) − 3$

4 _____

5 Solve the given system of equations by substitution. $y = x$
$x + y = 18$

A $(0, 0)$

B $(14, 4)$

C $(10, 8)$

D $(9, 9)$

5 _____

6 Which of the following is the correct graphic representation of the linear function $f(x) = 2x + 1$?

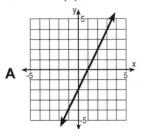
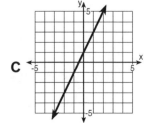
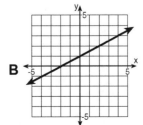
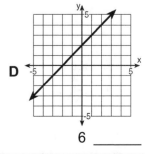

6 _____

7 What is the maximum number of times that the graph of a function can intersect the *y*-axis?

    A 3

    B 2

    C 1

    D 0
                              7 _____

8 Which of the following is the correct graphic representation of the linear function $f(x) = 3x - 4$?

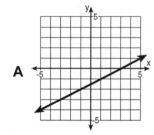 **A**
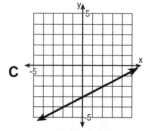 **C**

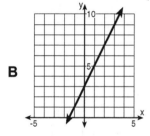 **B**
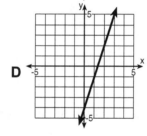 **D**

                                       8 _____

9 Two congruent angles that are supplementary must be _____ angles.

    A right

    B straight

    C acute

    D obtuse
                              9 _____

10 Rentals for a certain movie were graphed on the chart below.

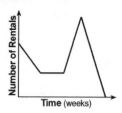

Which of the following statements best describes the number of rentals over the number of weeks?

    A Rentals were constant over the time period.

    B Rentals started at a maximum and decreased to zero.

    C Sales started high, decreased and leveled off, sharply increased to a peak, and quickly dropped to zero.

    D Rentals started high, decreased to a minimum, increased to a maximum, and fell to zero.
                                 10 _____

11 What is an equation of the linear function that represents the following table of values?

x	y
2	4
3	6
4	8

    A $f(x) = x - 2$

    B $f(x) = x + 2$

    C $f(x) = -2x$

    D $f(x) = 2x$
                                 11 _____

Test 1 – Part 1

12 The graph below indicates Juanita's distance from home as she travels to school.

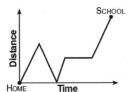

Which of the following best describes Juanita's journey to school?

A She leaves for school, stops to play with a puppy, continues on her way, stops at the market to buy a bottle of juice, and arrives at school.

B She leaves for school, returns home to retrieve a homework assignment, leaves for school again, waits for a train to pass, and arrives at school.

C She runs out the door, slows down to pick up a dollar, starts running again, stops to buy some candy, and runs to school.

D She dashes out the door, slows her pace down, starts skipping, jogs at a steady pace, and sprints to school.   12 _____

13 The area of a storeroom with a rectangular floor is 2,800 square yards. The perimeter of the storeroom floor is 220 yards. Which system of equations will determine the length (ℓ) and the width (w) of the storeroom floor?

A $\ell + w = 220$ and $\ell^2 + w^2 = 2,800$

B $\ell w = 2,800$ and $2\ell + 2w = 220$

C $\ell w = 2,800$ and $2\ell + w = 220$

D $\ell w = 220$ and $2\ell + 2w = 2,800$ 13 _____

14 $3^4 \cdot 3^7 =$

A 11^3

B 3^3

C 3^{11}

D 9^{11}

14 _____

15 What is the value of $\dfrac{6.3 \times 10^8}{3 \times 10^4}$ in scientific notation?

A 2.1×10^4

B 2.1×10^{-2}

C 2.1×10^{-4}

D 2.1×10^2 15 _____

16 Two congruent angles are supplementary. The measure of each angle is _____.

A $90°$

B $45°$

C $180°$

D $60°$ 16 _____

17 What is an equation of a linear function that represents the following table of values?

x	y
3	4
4	5
5	6

A $f(x) = 2x$

B $f(x) = 1 - x$

C $f(x) = x - 1$

D $f(x) = x + 1$ 17 _____

18 $5^2 \cdot 5 =$

    A 25^2

    B 25^3

    C 5^3

    D 5^2 18 _____

19 What is 25,000,000 written in scientific notation?

    A 25×10^6

    B 2.5×10^6

    C 25×10^7

    D 2.5×10^7 19 _____

20 Convert the given expression into scientific notation: $\dfrac{(5 \times 10^9)(8 \times 10^{-2})}{2 \times 10^{-3}} =$

    A 2×10^6

    B 2×10^9

    C 2×10^5

    D 2×10^{11} 20 _____

21 The cost of traveling an interstate highway is given by the table. What is an equation that can represent this relationship between cost (c) and the number of miles traveled (t)?

Miles	Cost
10	1.46
15	2.16
20	2.86
30	4.26

    A $c = 0.14t + 0.06$

    B $c = 0.14t + 6$

    C $c = 0.15t$

    D $c = 0.10t + 0.10$ 21 _____

22 A star is 3.4×10^8 light years away. In scientific notation, how far away (in light years) is a star that is half the distance?

    A 1.7×10^8

    B 1.7×5^8

    C 3.4×10^4

    D 1.7×10^4 22_____

23 $7^2 \cdot 7^8 =$

    A 7^{16}

    B 49^{16}

    C 7^{10}

    D 49^{10} 23 _____

Test 1 – Part 1

24 The table below shows the average amount of milk consumed per teenager between the years 1988 and 2003.

Consumption of Milk Per Teenager

Year	1988	1993	1997	1998	1999	2000	2001	2002	2003
Milk Consumption (gallons)	1.8	3.4	5.2	6.5	6.5	7.6	8.4	9.6	10.8

Which of the following graphed lines best models the data?

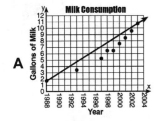

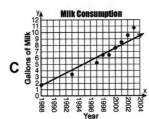

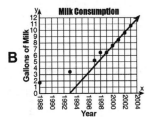

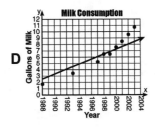

24_____

25 Irina uses data that she collected on how many minutes she and her friends studied for a test (x) and the grade each received on the test (y). She then determined the equation of best fit to be $y = 0.58x + 44.7$. Use this equation to predict approximately what grade a friend of hers would receive after studying for 75 minutes.

A 88

B 44

C 89

D 52

25 _____

26 What geometric shapes can be drawn from two sets of equivalent parallel sides measuring 3 inches and 4.5 inches?

A A trapezoid and a parallelogram

B A rectangle and a parallelogram

C A rectangle, a parallelogram, and a square

D A trapezoid, a parallelogram, and a rectangle

26_____

27 If two angles of a triangle measure 43° and 48°, the triangle is

A isosceles

B acute

C obtuse

D right

27 _____

28 Triangle ABC is congruent to triangle PQR. Which sequence of congruency transformations maps triangle ABC onto triangle PQR?

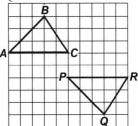

A a rotation of 90° clockwise about A followed by a translation

B a translation followed by a line reflection

C a point reflection followed by a translation

D a rotation of 90° counterclockwise about A followed by a line reflection

28 _____

Test 1 – Part 1

29 Which one of the following graphs shows the relationship "y is proportional to x"?

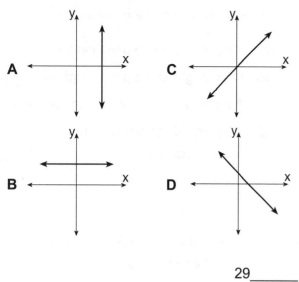

A

B

C

D

29 _____

30 The net of a cylinder is shown below.

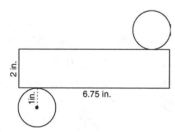

2 in.

1 in.

6.75 in.

What is the surface area of the cylinder to the *nearest square inch*?

A 16 in.²

B 19 in.²

C 13 in.²

D 6 in.²

30 _____

31 In the accompanying diagram, ΔA′B′C′ is the image of ΔABC.

What type of transformation is shown?

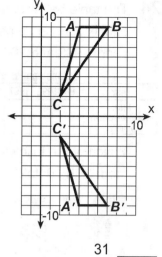

A dilation

B rotation

C translation

D reflection

31 _____

32 If the perimeter of an equilateral triangle is 36 inches, what is the perimeter of its image under a dilation with a scale factor of 4?

A 144 in.

B 32 in.

C 8 in.

D 72 in.

32 _____

33 Which of the following properties of an object are not preserved under a line reflection?

A shape

B size

C all of these

D orientation

33 _____

34 What is the area of the largest square graphed below?

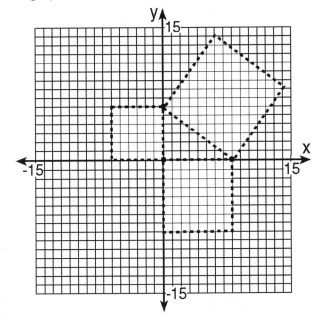

A 10 units^2

B 36 units^2

C 100 units^2

D 64 units^2 34 _____

35 The graph of the line passing through the points (6, 7) and (4, 2) has a slope of

A $\frac{2}{5}$

B $-\frac{5}{2}$

C $-\frac{1}{2}$

D $\frac{5}{2}$ 35 _____

36 Two lines have the following equations:

$$2x + y = 4$$
$$3y = 2x - 12$$

At what point do these lines intersect?

A (1, 2)

B (3, 10)

C (6, –8)

D (3, –2)   36 _____

37 In the lake shown below, an island is located at (5, 6). A boat travels in a straight line from (−3, −1) to the island.

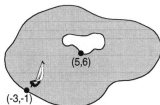

How far does the boat travel? Round the answer to the *nearest tenth* of a unit.

A 10.6 units

B 7.8 units

C 6.7 units

D 5.4 units   37 _____

38 Determine the equation of the line in the given graph.

A $x = 4$

B $y = 4x$

C $x = 4y$

D $y = 4$ 38 _____

39 Solve the equation for the given variable.

$$-23 = -16 + 8x$$

A $-\dfrac{1}{8}$

B $-\dfrac{7}{8}$

C $\dfrac{7}{8}$

D $\dfrac{1}{8}$

39 _____

40 $3z + 26 - 2z = -6$

A -4

B -32

C -20

D -16

40 _____

41 What is the solution of the system of equations whose graphs are shown below?

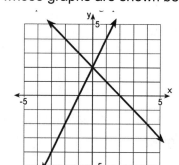

A $(1, 2)$

B $(2, 0)$

C $(0, 2)$

D $(2, -1)$

41 _____

42 Which of the following two-dimensional faces can be formed from a cross-section of a cube?

A rectangle, only

B square and rectangle, only

C square, rectangle, and hexagon

D square, only

42 _____

43 Which of the following properties of an object are preserved under a translation?

A size, shape, and orientation

B size and shape, only

C shape and orientation, only

D shape, only

43 _____

44 Triangle *ABC* is congruent to triangle *PQR*.

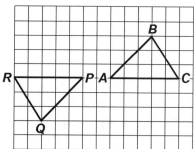

Which sequence of congruency transformations maps triangle *ABC* onto triangle *PQR*?

A a line reflection followed by a translation

B a rotation of 90° clockwise about *A* followed by a translation

C a point reflection followed by a translation

D a rotation of 90° counterclockwise about *A* followed by a line reflection

44 _____

Test 1 – Part 1

45 Keisha left home to walk two blocks to the school bus stop. After walking one block, she realized that she had left her soccer uniform at home. She ran home to get it and ran back to catch the bus. She only waited for one minute for the bus to arrive, then she rode the rest of the way to school. The graph of which function below correctly depicts this situation?

A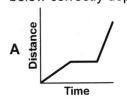

C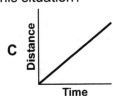

B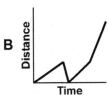

D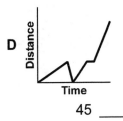

45 _____

46 The net of a cylinder is shown to the right.

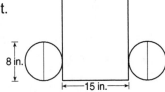

8 in.

|←——15 in.——→|

What is the lateral surface area of this cylinder to the *nearest square inch*?

A 189 in.²

B 377 in.²

C 478 in.²

D 754 in.²

45 _____

47 If the *y*-intercept of a line is 4, what could be an equation for this line?

A $y = x - 4$

B $y = 4x$

C $y = 4x + 3$

D $y = \frac{1}{2}x + 4$

47 _____

48 Triangle *ABC* is congruent to triangle *PQR*.

Which sequence of congruency transformations maps triangle *ABC* onto triangle *PQR*?

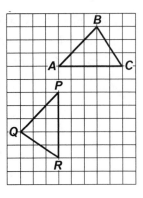

A a rotation of 90° counterclockwise about *A* followed by a translation

B a point reflection followed by a translation

C a rotation of 90° clockwise about *A* followed by a translation

D a rotation of 90° counterclockwise about *A* followed by a line reflection   48 _____

49 Use the accompanying graph to determine the unit rate of change of wages earned for time worked.

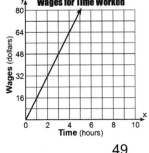

A $32 per hour

B $16 per hour

C $8 per hour

D $24 per hour

49 _____

50 The dimensions of a cylinder are shown below. What is the total surface area of this cylinder? Round the answer to the *nearest square inch*.

A 264 in.²

B 754 in.²

C 1,583 in.²

D 504 in.²

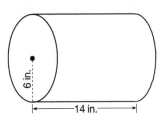

6 in.

|←——14 in.——→|

50 _____

51 As shown on the accompanying graph, what is the solution of the system of equations $y = 2x + 1$ and $x + y = 1$?

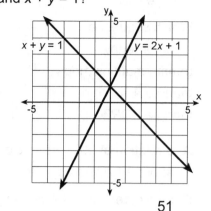

A (0, 1)

B (1, 0)

C (0, −1)

D (−1, 0)

51 _____

52 James uses data that he collected in a science experiment to calculate a line of best fit. He determines the equation of the line to be $y = 7x + 2.25$.

Use this equation to calculate the value of y when $x = 6$.

A 42

B 39.75

C 44.25

D 15.25

52 _____

53 What equation represents the missing step of the solution below?

(1) $3(x - 14) = 274 - x$

(2) $3x - 42 = 274 - x$

(3)   [   ]

(4) $x = 79$

A $2x = 316$

B $4x = 316$

C $4x = 232$

D $2x = 232$

53 _____

54 Mr. Adams drove his son up Route 11W to track practice. After watching practice for awhile, he went farther up Route 11W to the supermarket to buy groceries. When he returned to pick up his son, practice was nearly over, so he waited briefly, and then drove home. The graph of which function below correctly depicts this situation?

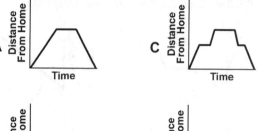

54 _____

55 In the accompanying diagram, $\triangle A'B'C'$ is the image of $\triangle ABC$.

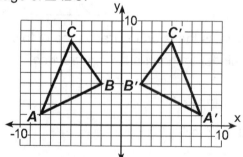

What type of transformation is shown in the illustration?

A rotation

B dilation

C translation

D line reflection

55 _____

Test 1 – Part 1

56 Solve and check: $\frac{3}{8}x - 6 = -24$

57 On the graph below, graph and label the function $f(x) = -x + 5$.

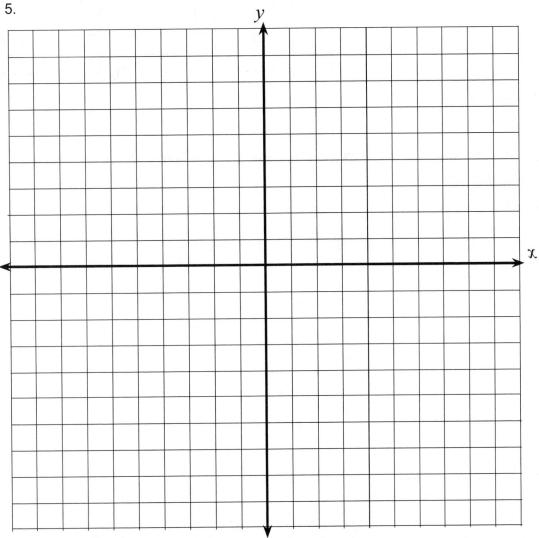

58 **Part A**

Graph the following system of equations on the coordinate plane below.

$y = 2x - 4$

$y = -x - 1$

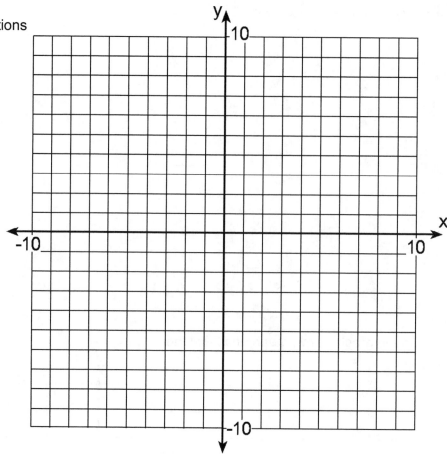

Part B

What is the solution to the system of equations graphed in Part A?

Show your work:

*Answer:*_____

59 A soccer field is 70 yards shorter than 3 times its width. Its perimeter is 380 yards. Find the length and width of the field.

Show your work:

*Answer:*_____

60 Solve and check: $48 = \frac{13}{7}x + 9$

61 Is the graph of the following points a function? *Justify your answer.*
$(-6, 4)$, $(-5, 4)$, $(-4, 3)$, $(-2, 3)$

62 Eric is working as an insurance salesperson. He earns a base salary and a commission on each new policy that he sells during the week. Last week he sold 3 new policies and earned $840 for the week. The week before, he sold 7 new policies and earned $1,220.

Part A

Find an linear function for Eric's weekly salary.

Part B

Interpret the rate of change and initial value for this situation.

63 Triangle *ABC* has coordinates *A*(1, 2), *B*(0, 5), and *C*(5, 4).

Part A

On the graph below, draw and label △*ABC*.

Part B

Graph and state the coordinates of △*A'B'C'*, the image of △*ABC* after the translation which maps (*x*, *y*) to (*x* − 6, *y* + 3).

Part C

Graph and state the coordinates of △*A"B"C"* the image of △*A'B'C'* after a reflection in the *x*-axis.

Part D

Graph and state the coordinates of △*ABC*, the image of △*A"B"C"* after a reflection in the origin.

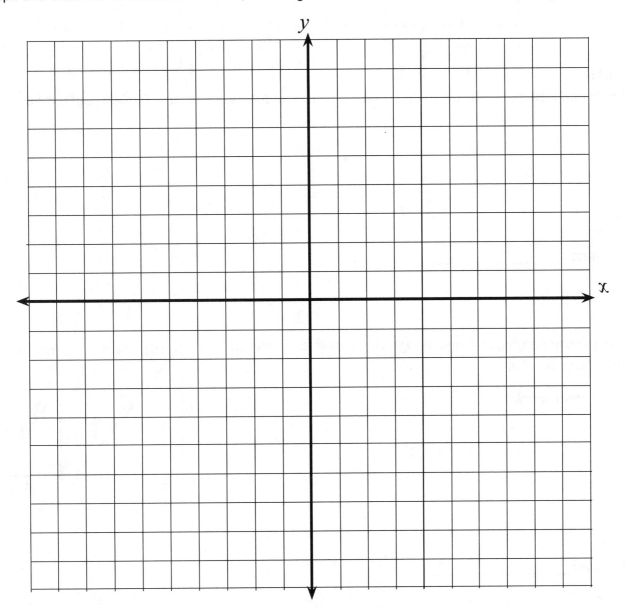

64 Is it possible for a rotation and a dilation to be equivalent?

Part A

Consider a rotation of 180° about the origin and a dilation of −1 centered at the origin.

Each graph below starts with same △*ABC*. On the first graph, rotate △*ABC* 180° about the origin. On the second graph, dilate △*ABC* with a SF = −1 centered at the origin.

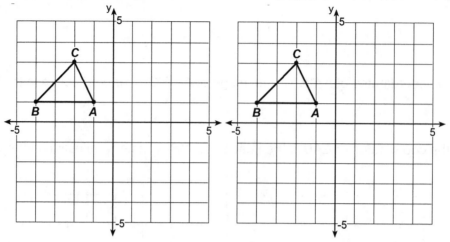

Part B

Are the images formed by the rotation and the dilation of △*ABC* equivalent? Explain your answer.

*Answer:*_____

65 In the accompanying diagram, triangle *A* is similar to triangle *B*. Find the value of *n*.

Show your work:

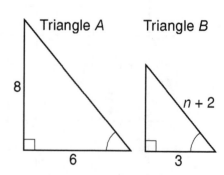

*Answer:*_____

TEST 2

1 What are the solutions to the equation $25x^2 = 4$?

A $x = \frac{5}{2}$, only

B $x = \pm\frac{2}{5}$

C $x = \frac{2}{5}$, only

D $x = \pm\frac{5}{2}$

1 _____

2 Express the given number or expression in scientific notation: 5,120,000

A 5.12×10^{-6}

B 51.2×10^5

C 0.512×10^6

D 5.12×10^6

2 _____

3 A star is 3.4×10^8 light years away. In scientific notation, how far away (in light years) is a star that is half the distance?

A 1.7×10^4

B 1.7×5^8

C 3.4×10^4

D 1.7×10^8

3 _____

4 Which of the following equations is never true?

A $5 - (x + 4) = x - 1$

B $5 + (x + 4) = x + 1$

C $5 - (x + 4) = 1 - x$

D $5 - (x + 4) = x + 1$

4 _____

5 Solve the given system of equations using addition or subtraction.

$$x + y = 8$$
$$x - y = 2$$

A $(5, 3)$

B $(3, 5)$

C $(3, 1)$

D $(10, -2)$

5 _____

6 Which of the following is not a linear function?

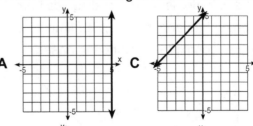
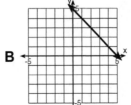
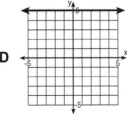

6 _____

7 Which of the following expressions can be used to find the values of $f(x)$ in the table below?

x	1	2	3	4	5	6	7
f(x)	5	7	9	11	13	?	?

A $f(x) = 3x + 2$

B $f(x) = x + 4$

C $f(x) = 2x + 3$

D $f(x) = 3x$

7 _____

8 Which of the following is the correct graphic representation of the function $f(x) = -3$?

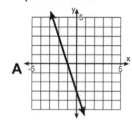 A

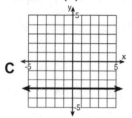 C

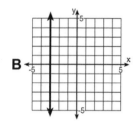 B

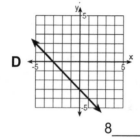 D

8 _____

9 Antoinette has just hit a home run. Which of the following graphs best represents the relationship between her distance from the home plate and the length of time it takes her to run completely around the bases?

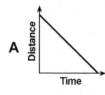 A

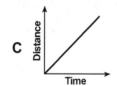 C

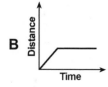 B

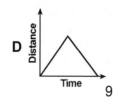 D

9 _____

10 What is an equation of the linear function whose graph passes through the points $(-4, 13)$ and $(8, 4)$?

A $f(x) = -\frac{4}{3}x - 10$

B $f(x) = \frac{3}{4}x - 10$

C $f(x) = \frac{4}{3}x + 10$

D $f(x) = -\frac{3}{4}x + 10$

10 _____

11 Roy is playing on a swing in his backyard. Which of the following graphs best models the relationship between his height (h) above ground when he is swinging and the amount of time (t) he spends swinging?

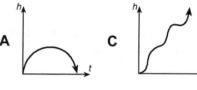

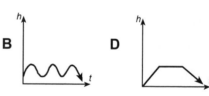

11 _____

12 The length of a rectangle is 5 meters less than twice its width. If the perimeter of the rectangle is 38 meters, then what is the width?

A 8 m

B 11 m

C 7 m

D 5 m

12 _____

Test 2 – Part 1

13 10^1 x 10^1 is equal to _____.

A 10

B 20

C 10^2

D 10^0 13 ____

14 Convert the given expression into scientific notation: $\dfrac{(15 \times 10^8)(4 \times 10^5)}{3 \times 10^4} =$

A 2×10^{11}

B 2×10^{10}

C 2×10^{17}

D 2×10^7 14 ____

15 Two congruent angles that are complementary must be _____ angles.

A acute

B straight

C right

D obtuse   15 ____

16 What is an equation of the linear function that represents the following table of values?

x	y
2	1
4	2
6	3

A $y = x - 2$

B $y = x - 1$

C $y = \dfrac{1}{2}x$

D $y = 2x$ 16 _____

17 The expression $3^2 \cdot 3^3 \cdot 3^4$ is equivalent to

A 3^9

B 27^9

C 3^{24}

D 27^{24} 17 _____

18 Expressed in decimal notation, 4.726×10^{-3} is

A 0.004726

B 4,726

C 0.04726

D 472.6   18 _____

19 The mass of a microscopic organism is approximately 1.0×10^{-14} grams. How many of these organisms are needed to make up 1 gram?

A 10^{15}

B 10^{14}

C 1,400

D 10^{28} 19 ____

20 Ava recorded the number of hours she made waiting tables (x) over ten days and how much money (y) she earned each day. She then calculated the line of best fit using this data. If the line of best fit is $y = 8.5x + 2.25$, determine how many hours (to the nearest tenth of an hour) she would have to work to make $90.

A 10.9

B 767.3

C 8.4

D 10.3   20 _____

21 A survey of high school students asked, "What is your favorite school lunch?". The results are shown in the two-way frequency table below.

'Favorite School Lunch' Poll

Grade	Lunch Choice				
	PIZZA	TACOS	CHICKEN	OTHER	TOTAL
9th	53	21	13	23	110
10th	41	11	9	44	105
11th	61	23	16	15	115
12th	38	29	9	19	95
TOTAL	193	84	47	101	425

In reference to the total number of participants in the survey, what is the relative frequency of a 9th grade student response?

A 0.275

B 0.259

C 0.209

D 0.454   21 _____

22 Use the data in the accompanying table to describe the slope of the line of best fit.

x	-2	-2	1	1	5	5
y	2	-1	2	1	2	1

A negative slope

B undefined slope

C zero positive slope

D positive slope   22 _____

23 The two-way frequency table below shows the number of hours math students spent on homework and whether they worked by themselves or had help completing the assignment.

Math Homework

	Worked One Hour or Less	Worked More than One Hour
Worked Alone	45	25
Worked with Help	20	35

What is the relative frequency of math students who had help doing their homework and worked more than 1 hour in reference to all the students who received help on their homework?

A 0.58

B 0.364

C 0.28

D 0.636   23 _____

Test 2 – Part 1

24 Lance is drawing a triangle. He draws one side that is 8 inches long and another side that is 5 inches long. Which one of the following could be the third side?

A 1 inch

B 3 inches

C 4 inches

D 2 inches   24 _____

25 In the accompanying diagram of △ABC, side $\overline{BC}$ is extended to D, m∠B = 2y°, m∠BCA = 6y°, and m∠ACD = 3y°. What is m∠A?

A 20°

B 24°

C 17°

D 15°

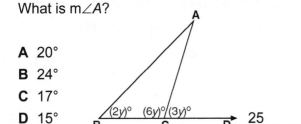

25 _____

26 What is the value of x for which the equation $9^{-6} \cdot 9^x = 1$ is true?

A 0

B −6

C $\frac{1}{6}$

D 6   26 _____

27 Triangle ABC is congruent to triangle PQR.

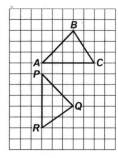

Which sequence of congruency transformations maps triangle ABC onto triangle PQR?

A a rotation of 90° clockwise about A followed by a translation

B a rotation of 90° counterclockwise about A followed by a line reflection

C a line reflection followed by a translation

D a point reflection followed by a line reflection   27 _____

28 In the accompanying diagram, line $\overleftrightarrow{LM}$ and line $\overleftrightarrow{OP}$ intersect at point S.

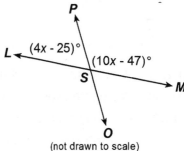

(not drawn to scale)

If m∠LSP = (4x − 25)° and m∠PSM = (10x − 47)°, find m∠LSP.

A 18°

B 47°

C 133°

D 36°   28 _____

29 Which one of the following tables is an example of y being proportional to x?

A
x	y
2	4
3	5
4	6

C
x	y
2	4
3	3
4	2

B
x	y
2	4
3	9
4	16

D
x	y
2	4
3	6
4	8

29 _____

30 What is the surface area of a cylinder with a radius of 24 cm and a height of 14 cm? Round your answer to the *nearest hundredth* of a square centimeter.

A 2,110.08 cm²

B 5,727.36 cm²

C 5,260.80 cm²

D 2,863.68 cm²   30 _____

31 When a copy machine makes a copy, the image (copy) is a dilated version of the original. If Willa makes a copy on a copy machine and the copy is the same size as the original, what is the scale factor of the dilation?

A 2

B 0

C 1

D $\frac{1}{2}$ 31 _____

32 The graph of $y = x^3$ is shown below.

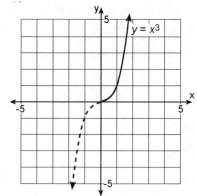

What two reflections would be required to map the solid line part of the graph onto the dotted line part?

A a reflection over the line $y = x$ followed by a reflection over the x-axis

B a reflection over the line $y = x$ followed by a reflection over the y-axis

C a reflection over the x-axis followed by a reflection over the y-axis

D none of these   32 _____

33 Triangle $A'B'C'$ is the image of $\triangle ABC$ under a dilation such that $A'B' = \frac{1}{2} AB$. Triangles ABC and $A'B'C'$ are _____.

A neither congruent nor similar

B congruent, but not similar

C both congruent and similar

D similar, but not congruent   33 _____

34 The cube shown in the accompanying diagram has sides of length 8. What is the length of the diagonal *AB*?

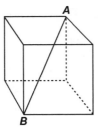

A $\sqrt{182}$

B 192

C 24

D $\sqrt{192}$

34 _____

35 What is the slope of a line that passes through points (−4, 2) and (6, 8)?

A $-\frac{3}{5}$

B $\frac{3}{5}$

C $\frac{5}{3}$

D $-\frac{5}{3}$

35 _____

36 The graphs of the equations in the system of equations $3x + y = 6$ and $y + 3x = 6$ will

A not intersect

B intersect at all points on the line $y = -3x + 6$

C will intersect at exactly two points

D intersect at exactly one point

36 _____

37 What is the length of the radius of a circle whose center is at (6, 0) and passes through (2, −3)?

A 11

B 4

C 5

D 7

37 _____

38 The slope of a straight line is always

A increasing

B decreasing

C constant

D positive

38 _____

39 The steps for solving the equation $3(2x - 6) = 2(3x - 9)$ are shown below.

(1) $3(2x - 6) = 2(3x - 9)$

(2) $6x - 18 = 6x - 18$

(3) $6x - 6x - 18 + 18 = 6x - 6x - 18 + 18$

(4) $0 = 0$

What is the correct conclusion?

A The equation is true for all values of *x*.

B The solution set is the empty set.

C $x = 18$ is the only solution.

D $x = 0$ is the only solution.

39 _____

40 Solve the equation for the given variable.

$$2 + 7y = 11 - y - y$$

A 1

B 13

C −1

D 3   40 ____

41 What is the solution to the system of equations graphed below?

A (−3, 3)

B (−3, −3)

C (3, −3)

D (3, 3)   41 ____

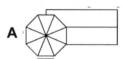

42 Which of the following properties of an object are not preserved under a rotation?

A orientation

B size

C shape

D none of these   42 ____

43 The drawing below shows a view of a barn with an attached silo.

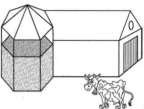

Which of the following drawings best represents the top view of this building?

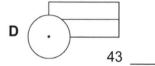

43 ____

44 Triangle *ABC* is congruent to triangle *PQR*.

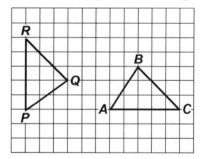

Which sequence of congruency transformations maps triangle *ABC* onto triangle *PQR*?

A a point reflection in *B* followed by a translation

B a rotation of 180° clockwise about *A* followed by a translation

C a rotation of 90° counterclockwise about *A* followed by a line reflection

D a rotation of 90° clockwise about *A* followed by a line reflection   44 ____

45 The rate at which water is entering a water tank, for any time $t > 0$ is shown in the graph below. A positive rate indicates that water is flowing into the tank, and a negative rate indicates that water is leaking from the tank.

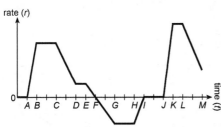

What is the longest interval of time for which the volume of water in the tank is increasing?

A *AF*

B *IK*

C *OF*

D *AB*

45 _____

46 The net of a cylinder is shown.

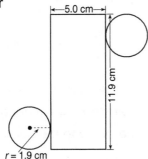

What is the total surface area of this cylinder rounded to the nearest whole number?

A 60 cm²

B 82 cm²

C 23 cm²

D 71 cm²

46 _____

47 If point *P* below was reflected over the *x*-axis, what are the coordinates of *P'*?

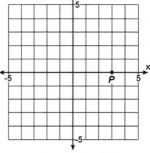

A $(-3, 0)$

B $(0, -3)$

C $(3, 0)$

D $(0, 3)$

47 _____

48 Which of the following is a graphic representation of a proportional relationship?

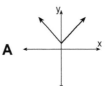

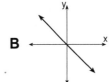
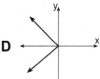

48 _____

49 What is the solution to the system of equations graphed?

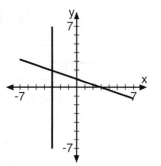

A $(2, -3)$

B $(-3, 2)$

C $(-2, 3)$

D $(3, 2)$

49 _____

50 Determine the equation of the line in the given graph.

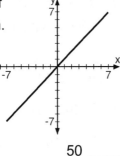

A $x = -y$

B $y = x$

C $-y = x$

D $y = -x$ 50 ____

51 What equation represents the second step of the solution shown below?

(1) $3(x - 2) = 5(x - 3) + 2$

(2) _____

(3) $7 = 2x$

(4) $3.5 = x$

A $3x - 6 = 5x - 6$

B $3x - 2 = 5x - 1$

C $3x - 6 = 5x - 1$

D $3x - 6 = 5x - 13$ 51 ____

52 A survey of students asked, "How many children are in your family?". The results are shown in the table below.

'Number of Children in Family' Poll

		Number of Children in Family					
		1	2	3	4	5	6	TOTAL
School Grade	7th	21	40	32	30	2	1	126
	8th	23	31	39	31	1	2	127
	9th	22	30	32	36	2	1	123
	TOTAL	66	101	103	97	5	4	376

Based on the survey results shown, what is the relative frequency that an 8th grade student has either 2 or 3 children in his or her family?

A 0.343

B 0.186

C 0.551

D $0.3\overline{3}$ 52 ____

53 Snow is falling at a rate of one foot an hour on top of a base 3 feet deep. Which of the following graphs best represents the relationship between the depth of the snow on the ground and the amount of time that has passed?

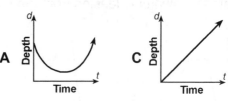

53 ____

54 The formula for finding the volume of a cylinder is

A $V = \pi r^2 h$

B $V = \frac{1}{3}\pi r^2 h$

C $V = 2\pi r^2 h$

D $V = \frac{4}{3}\pi r^3$ 54 ____

55 In the accompanying diagram, which transformation changes the solid-line parabola to the dotted-line parabola?

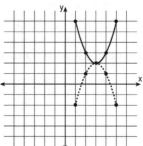

A line reflection, only

B translation

C line reflection or rotation

D rotation, only 55 ____

Test 2 – Part 1

56 One side of a triangle is 9 cm more than half as long as the longest side of the triangle. The third side is 8 cm shorter than the longest side. Find the length of each side if the perimeter is 181 cm.

Show your work:

*Answer:*_____

57 Fill in the accompanying table with the circumference and area of a circle using the given various lengths of the radius. Express all output value in terms of π.

Radius Length of a Circle	1	3	4	5	7	8	10	12	15
Circumference of a Circle									
Area of a Circle									

Does either the circumference or area row of the table represent a linear function? If yes, write an equation for this linear function.

58 **Part A**

Graph the system of equations on the coordinate plane provided.

$y = \frac{1}{2}x + 4$

$y = -\frac{1}{2}x - 4$

Part B

What is the solution to the system graphed in Part B?

Show your work:

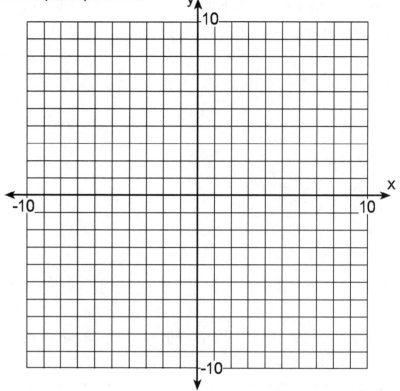

Answer:_____

59 One side of a triangular garden is 4 m longer than the shortest side of the garden. This side is also 5 m shorter than the longest side of the garden. If the perimeter of the garden is 46 meters, find the length of each side of the garden.

Show your work:

Answer:_____

60 Jill is three times as old as Nancy. If the sum of their ages is 52, find the age of each.

61 **Part A**

Complete the accompanying table
for the function $f(x) = -2x - 3$.

x	$f(x)$
-2	
-1	
	-3
	-5
2	

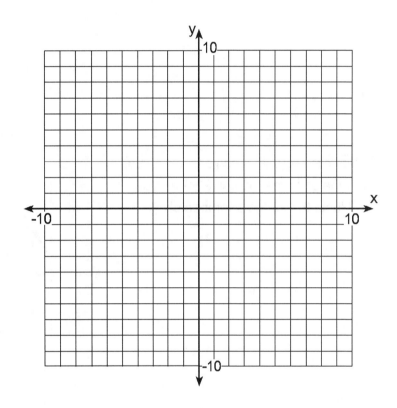

Part B

Using the accompanying graph and the
input and output values from the table in
Part A, graph the function $f(x) = -2x - 3$.

62 The first of two lines passes through the points (−2, −13) and (4, 2). The second line passes through the points (8, 12) and (2, −3). Determine algebraically whether these two lines intersect, are parallel, or are coinciding lines. If the lines intersect, state the coordinates of the point they have in common.

Show your work:

*Answer:*_____

63 Rotate triangle *RST* 90° clockwise about point *T* and label the image appropriately.

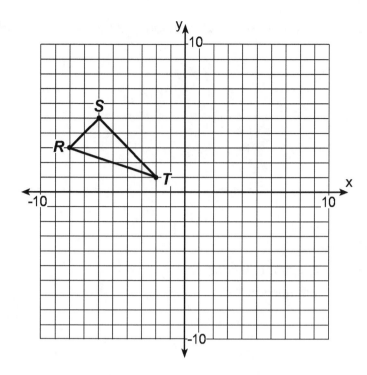

64 Dilate circle *O* with a scale factor of 1.5. Use the origin as the center of dilation and label the image appropriately.

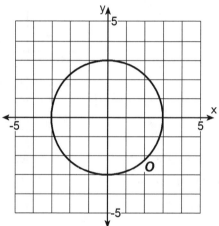

65 A caterer charges $22 per person for providing food and services at a graduation party. On the axes provided, sketch a graph that represents this proportional relationship for a party of a maximum of 10 guests.

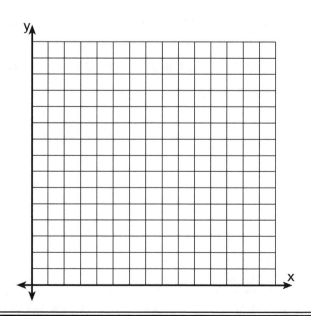

Test 2 – Part 2

TEST 3

1 $5^6 \cdot 5^3 =$

    A 5^9

    B 18^5

    C 25^9

    D 5^{18} 1 _____

2 What is the value of x for which the equation $8^{-2} \cdot 8^x = \dfrac{1}{512}$ is true?

    A -1

    B -2

    C 2

    D 1 2 _____

3 What is the value of $\sqrt{0.09}$?

    A 3

    B 1.8

    C 0.45

    D 0.3 3 _____

4 How is the number 192.57×10^{-6} expressed in standard form?

    A 0.0019257

    B 0.00019257

    C 0.019257

    D $192,570,000$ 4 _____

5 The two acute angles in an isosceles right triangle must measure

    A $30°$ and $60°$

    B $35°$ and $55°$

    C $45°$ and $45°$

    D $40°$ and $50°$ 5 _____

6 Convert the given expression into scientific notation: $\dfrac{\left(5 \times 10^9\right)\left(8 \times 10^{-2}\right)}{2 \times 10^3}$

    A 2×10^{11}

    B 2×10^5

    C 2×10^9

    D 2×10^8 6 _____

7   The measures of two angles of a triangle are 70° and 55°. This triangle is

    A   a right triangle

    B   an isosceles triangle

    C   an equilateral triangle

    D   an obtuse triangle   7 _____

8   Which one of the following tables represents a proportional relationship?

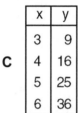

x	y
3	2
4	3
5	4
6	5

A

x	y
3	9
4	16
5	25
6	36

C

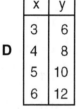

x	y
3	4
4	5
5	6
6	7

B

x	y
3	6
4	8
5	10
6	12

D

        8 _____

9   What equation represents the third step of the solution below?

    (1) $2(x + 30) = 180 - x$

    (2) $2x + 60 = 180 - x$

    (3)   [   ]

    (4) $x = 40$

    A $2x = 240$

    B $3x = 120$

    C $3x = 240$

    D $x = 120$ 9 _____

10   What is the solution to the system of equations graphed?

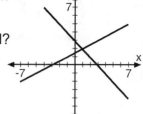

    A   (2, 1)

    B   (1, 2)

    C   (2, −1)

    D   (−1, 2)   10 _____

11   In 2011, the population of New York City was 8.24491×10^6. If the land area of New York City is 468.5 square miles, then what is the population density (people per square mile)?

    A   approximately 175,985 people per square mile

    B   approximately 1.7599×10^3 people per square mile

    C   approximately 17,599 people per square mile

    D   approximately 1.75985×10^6 people per square mile   11_____

12   Use the data in the accompanying table to describe the slope of the line of best fit.

x	-7	-3	-1	1	5	9
y	-8	-2	0	3	6	11

    A   zero slope

    B   negative slope

    C   undefined slope

    D   positive slope   12 _____

13 The United States federal budget for 2012 contained total expenditures of $3,796,000,000,000. If these expenditures were divided equally among all U.S. citizens, how much would it cost per citizen, on average, if the 2012 population is 3.14×10^8?

  A $121

  B $120,892

  C $1,209

  D $12,089   13 _____

14 In the fiscal year of 2012, total United States federal government spending was approximately 3.563 trillion dollars. If there were approximately 1.2×10^8 American households in existence during this year, then what was the average spending share of each household?

  A $2,969,167

  B $2,969

  C $29,692

  D $296,917   14 _____

15 What is a dilation scale factor that will produce an image congruent to the original?

  A   1

  B   2

  C   3

  D   0   15 _____

16 What is an equation of the linear function whose graph passes through the points (3, −17) and (−2, 8)?

  A $f(x) = -5x - 2$

  B $f(x) = -2x - \frac{1}{5}$

  C $f(x) = -2x - 5$

  D $f(x) = -\frac{1}{5}x - 2$ 16 _____

17 What is an equation of the linear function whose graph passes through the points (4, 39) and (10, 69)?

  A $f(x) = 5x + 19$

  B $f(x) = \frac{1}{5}x + 19$

  C $f(x) = 19x + 5$

  D $f(x) = 19x + \frac{1}{5}$ 17 _____

18 The graphs of the equations in the system of equations $4x - y = 6$ and $y - 4x = 7$ will

  A   intersect at exactly two points

  B   intersect at exactly one point

  C   not intersect

  D   intersect at all points on the line $y = 4x + 7$ 18 _____

Test 3 – Part 1

19 The cost of renting a car consists of a cost per day and a cost per kilometer. The cost of renting a car for 2 days and 150 km is $69. The cost of renting the same car for 3 days and 250 km is $108. What is the cost per day?

A $18.00

B $21.00

C $23.00

D $30.75   19_____

20 What is the solution to the system of equations graphed?

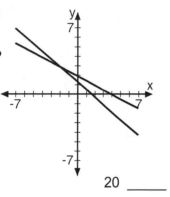

A (3, −2)

B (−3, 2)

C (2, −3)

D (−2, 3)   20 _____

21 To determine if a graph is a function, you can perform

A the horizontal line test

B the vertical line test

C the parallel line test

D the diagonal line test   21_____

22 Which of the following representations cannot be used to represent the function

$$f(x) = \{(2, 4), (3, 9), (4,16)\}?$$

A $y = x^2$ and the domain is {2, 3, 4}

B y is the square of x for the integers $2 \le x \le 4$

C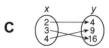

D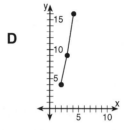

22 _____

23 Which of the following is the correct graphic representation of the function $f(x) = -\dfrac{1}{3}x + 4$?

A 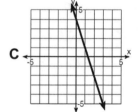   C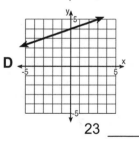

B 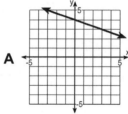   D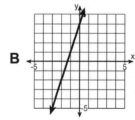

23 _____

24 In the accompanying diagram, line $\overleftrightarrow{AB}$ and line $\overleftrightarrow{CD}$ intersect at point E.

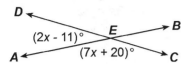

If $m\angle AEC = (7x + 20)°$ and $m\angle AED = (2x - 11)°$, find the value of x.

A   −6

B   15

C   23

D   19   24 _____

25 Which of the following is the correct graphic representation of the function $f(x) = \frac{1}{2}x + 3$?

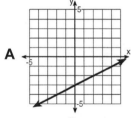

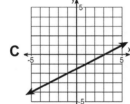

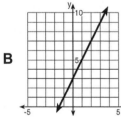

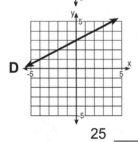

25 _____

26 What is the value of x for which the equation $4^x \div 4^{-3} = \frac{1}{16}$ is true?

A   1

B   −1

C   5

D   −5   26 _____

27 Which of the following is the correct graphic representation of the function $f(x) = \frac{1}{4}x - 3$?

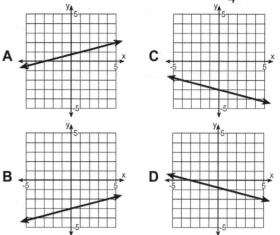

27 _____

28 What is an equation of the linear function that represents the following table of values?

x	y
0	-3
1	-2
2	-1

A $g(x) = x - 2$

B $g(x) = x - 3$

C $g(x) = -3x$

D $g(x) = -x + 1$ 28 _____

29 Triangle $A'B'C'$ is the image of $\triangle ABC$ under a dilation such that $A'B' = 3AB$. Triangles ABC and $A'B'C'$ are _____.

A   congruent, but not similar

B   neither congruent nor similar

C   similar, but not congruent

D   both congruent and similar   29 _____

Questions 30 and 31 refer to the following:

The rate at which water is entering a water tank, for any time $t > 0$, is shown in the accompanying graph. A positive rate indicates that water is flowing into the tank, and a negative rate indicates that water is leaking from the tank.

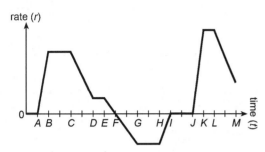

30 Over which interval(s) of time is the rate of water flow either in or out of the tank a constant?

    A   *OA* and *IJ*

    B   *BC* and *GH*, only

    C   *BC*, *DE*, *GH*, and *KL*

    D   *OM*   30 _____

31 Over which interval(s) of time is water neither flowing in nor out of the tank?

    A   *OA* and *IJ*

    B   *IJ*, only

    C   *OA*, only

    D   *OA*, *BC*, *DE*, *GH*, *IJ*, and *KL*   31 _____

32 Which of the following is not a rational number?

    A $3.\overline{63}$

    B $\sqrt{25}$

    C   2.4224222422224...

    D   7   32 _____

33 Solve the given system of equations by substitution. $2r + 5s = 10$
$$4r - 3s = -6$$

    A   (0, 2)

    B   (3, −6)

    C   (5, 2)

    D   (5, 0)   33 _____

34 If the sum of the measures of two angles of a triangle is equal to the measure of the third angle, the triangle must be

    A   right

    B   acute

    C   obtuse

    D   isosceles   34 _____

35 Which of the following represents the graph of the equation $x + 3y = 13$?

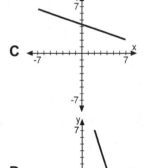

                              35 _____

36 The accompanying table shows the relative frequency, based on the total number of people surveyed, of individuals who were asked "Do you own a tablet computer?".

Do You Own a Tablet Computer?

	Owns a Tablet	Does Not Own a Tablet	TOTAL
Males		0.24	
Females	0.32		0.48
TOTAL			

In reference to the total number of participants in the survey, what is the relative frequency that an individual will own an tablet computer?

A 0.28

B 0.60

C 0.32

D 0.52   36 _____

37 Under a certain transformation, $\triangle A'B'C'$ is the image of $\triangle ABC$. The perimeter of $\triangle A'B'C'$ is twice the perimeter of $\triangle ABC$. This transformation is a _____.

A dilation

B translation

C reflection

D rotation   37 _____

38 Use the number line below to answer the given question.

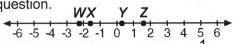

Which of the letters shown represents $\frac{1}{3}$ on the number line?

A *W*

B *Z*

C *Y*

D *X*   38 _____

39 In which of the following figures is $\triangle A'B'C'$ a reflection of $\triangle ABC$ in line ℓ?

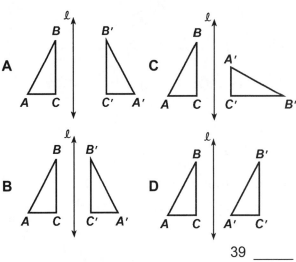

39 _____

40 Reflect triangle *QRS* over the *y*-axis and label the image appropriately.

What are the coordinates of point *R'*?

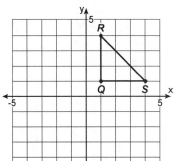

A (−4, 1)

B (−1, −4)

C (1, 4)

D (−1, 4)   40 _____

41 A triangle has coordinates $A(-1, -2)$, $B(-4, -2)$ and $C(-4, -5)$. What are the coordinates of point *A'*, the image of point *A*, under a dilation with a scale factor of 3?

A (2, 1)

B (−6, −3)

C (−12, −6)

D (−3, −6)   41 _____

42 In the figure below, $\overline{AB} \cong \overline{BC}$.

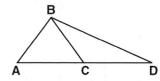

If m∠ABC = 72°, what is m∠BCD?

A   108°

B   72°

C   106°

D   126°   42 _____

43 Which of the following is the correct graphic representation of the function $f(x) = -\frac{3}{4}x + 2$?

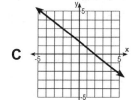

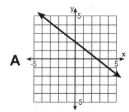

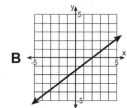

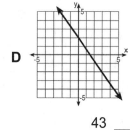

43 _____

44 A rectangular prism has dimensions of x by y by z. The length of the longest diagonal is

A $x^2 + y^2 + z^2$

B $\sqrt{x^2} + \sqrt{y^2} + \sqrt{z^2}$

C $\sqrt{x^2 + y^2} + \sqrt{x^2 + z^2} + \sqrt{y^2 + z^2}$

D $\sqrt{x^2 + y^2 + z^2}$ 44 _____

45 Which of the following is the correct graphic representation of the function $f(x) = \frac{3}{5}x - 1$?

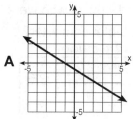

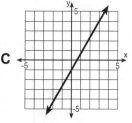

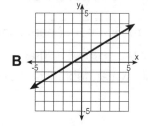

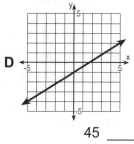

45 _____

46 If the coordinates of A are (3, 4) and the coordinates of B are (−3, −4), then the length of $\overline{AB}$ is

A   10

B   100

C   5

D   20   46 _____

47 Use the formula $V = \pi r^2 h$ to calculate the volume of a cylinder.

A grain silo has an inside diameter of 4.7 meters and a height of 10.4 meters. How many cubic meters of grain will the silo hold? Round your answer to the *nearest tenth*.

A   721.4 m³

B   180.3 m³

C   153.5 m³

D   399.1 m³   47 _____

48 Wind Whisper Kites sells plastic kites for $3.75 each. A table representing the number of kites purchased and the cost is shown below.

Number Purchased	Total Cost ($)
0	0
1	3.75
2	7.50
3	11.25
4	15.00
SEASON SPECIAL: BUY 4, GET 1 FREE	15.00

Which of the following images represents a scatter plot of the data in the table?

A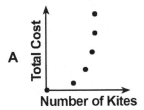

C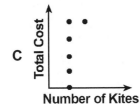

B

D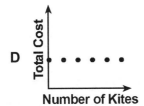

48 _____

49 The function, *f*, is represented by the accompanying graph.

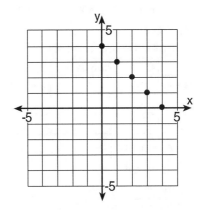

Which of the following representations is not equivalent to the function defined in this graph?

A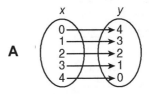

B The set of ordered pairs such that *x* is a whole number less than or equal to 4, and *y* is 4 more than the additive inverse of *x*.

C *y* = *x* + 4 with a domain of {−4, −3, −2, −2, 0}

D {(0, 4), (1, 3), (2, 2), (3, 1), (4, 0)}

49 _____

50 Mall shoppers were asked whether or not they owned a digital camera. The data from this survey is shown in the accompanying two-way frequency table.

Digital Camera Ownership

	Owner	Non-owner	TOTAL
Males	50	30	80
Females	45	25	70
TOTAL	95	55	150

A table showing the relative frequency by column for this data is

A

Digital Camera Ownership

	Owner	Non-owner	TOTAL
Males	0.53	0.55	0.53
Females	0.47	0.45	0.47
TOTAL	1.00	1.00	1.00

B

Digital Camera Ownership

	Owner	Non-owner	TOTAL
Males	0.33	0.20	0.53
Females	0.30	0.17	0.47
TOTAL	0.63	0.37	1.00

C

Digital Camera Ownership

	Owner	Non-owner	TOTAL
Males	0.67	0.40	0.60
Females	0.33	0.60	0.40
TOTAL	1.00	1.00	1.00

D

Digital Camera Ownership

	Owner	Non-owner	TOTAL
Males	0.33	0.53	0.53
Females	0.67	0.47	0.37
TOTAL	1.00	1.00	1.00

50 _____

51 What is the volume of the cylinder shown? Round your answer to the nearest cubic centimeter.

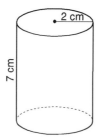

A   87 cm³

B   28 cm³

C   44 cm³

D   88 cm³

51 _____

52 A survey of high school students asked, "What is your favorite school lunch?". The results are shown in the two-way frequency table below.

'Favorite School Lunch' Poll

Grade	Lunch Choice				
	PIZZA	TACOS	CHICKEN	OTHER	TOTAL
9th	53	21	13	23	110
10th	41	11	9	44	105
11th	61	23	16	15	115
12th	38	29	9	19	95
TOTAL	193	84	47	101	425

Based on the data shown, what is the relative frequaency of tacos as an 11th grade student's favorite lunch?

A   0.53

B   0.20

C   0.054

D   0.274

52 _____

53 The graph shows Miguel's distance from school with respect to time. Which of the following statements best represents this graph?

A   Miguel started driving to school, but he got stuck in a traffic jam and sat in the same spot until he gave up and returned home.

B   Miguel had to stop at two red lights on his way to pick up his brother at soccer practice.

C   Miguel stopped once to talk to a friend on his cell phone on his way to the grocery store.

D   Miguel's speed varied between 10 mph and 30 mph depending upon the flow of traffic between his house and school.

53 _____

Test 3 – Part 1

54 Middle school students were asked whether or not they participate in athletics. The data from this survey is shown in the accompanying two-way frequency table.

Student Participation in Athletics

	Athlete	Non-athlete	TOTAL
Boys	50	10	60
Girls	25	15	40
TOTAL	75	25	100

A table showing the relative frequency by row for this data is

A

Student Participation in Athletics

	Athlete	Non-athlete	TOTAL
Boys	0.833	0.167	1.00
Girls	0.625	0.375	1.00
TOTAL	0.75	0.25	1.00

B

Student Participation in Athletics

	Athlete	Non-athlete	TOTAL
Boys	0.50	0.10	0.60
Girls	0.25	0.15	0.40
TOTAL	0.75	0.25	1.00

C

Student Participation in Athletics

	Athlete	Non-athlete	TOTAL
Boys	0.67	0.33	1.00
Girls	0.34	0.66	1.00
TOTAL	0.51	0.49	1.00

D

Student Participation in Athletics

	Athlete	Non-athlete	TOTAL
Boys	0.67	0.04	0.60
Girls	0.33	0.06	0.40
TOTAL	1.00	1.00	1.00

54 _____

55 A student performed an experiment by placing the numbers 0–5 on each of three different colored cards. The student placed the cards in a dark bag and randomly drew a card. The student noted the number and color and returned it to the bag. Use the results shown in the table as the sample space for the given question.

'Random Card Selection' Experiment

		Card Number						
		0	1	2	3	4	5	TOTAL
Card Color	Red	5	7	6	3	2	5	28
	Green	7	6	9	4	6	7	39
	Yellow	7	8	8	4	5	1	33
	TOTAL	19	21	23	11	13	13	100

If the card selected in the experiment described shows the number 4, what is the relative frequency that it is a yellow card?

A 0.05

B $0.\overline{15}$

C 0.385

D $0.\overline{11}$

55 _____

Test 3 – Part 1

56 Part A

Complete the table below for the function $h(x) = x - 2$.

x	h(x)
-2	
1	
3	

Part B

On the graph below, sketch a graph of this function and label the graph with its equation.

Part C

Using your graph for Part B, predict where the graph of $y = 2$ will intersect the graph of $h(x) = x - 2$.

Show your work:

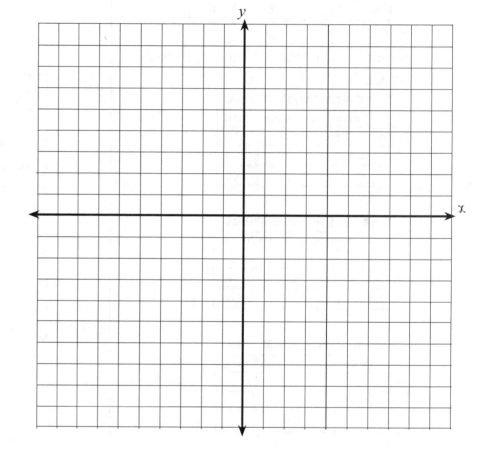

*Answer:*_____

57 Triangle $A'B'C'$ is the image of triangle ABC after a dilation.

Determine the coordinates of the center of dilation and the scale of dilation.

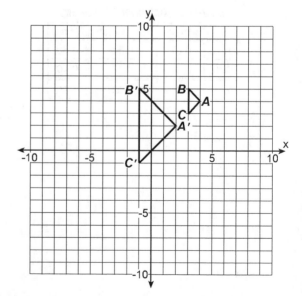

*Answer:*_____

58 Determine the solution to the system
of equations whose graph is shown:

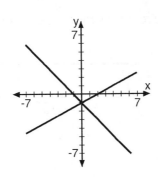

59 In the chart below, select values for *x* and determine the values for *f(x)* for the function $f(x) = x^2 + 3$.
Then graph the function and determine if it is linear or nonlinear.

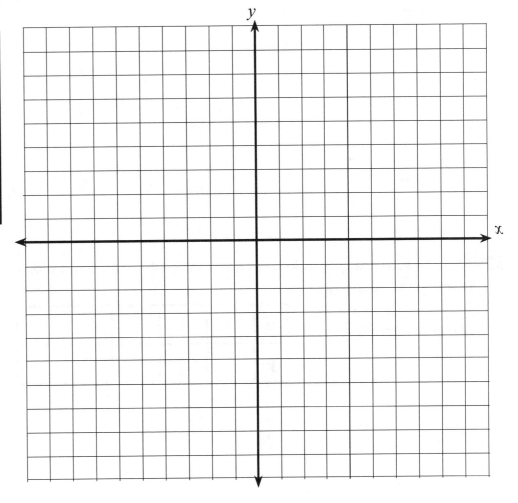

x	f(x)	(x,f(x))

60 Find the total surface of the right circular cylinder shown in the given diagram. Express your answer in terms of π.

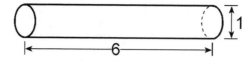

61 Solve for *x*: $40 - 0.05(x - 100) = 10$

62 Fill in the accompanying table with *x* and *y* values that represent a proportional relationship that has the indicated rate of change.

rate of change = 0.4

x					
y					

63 The three different linear functions shown below are represented by three different methods.

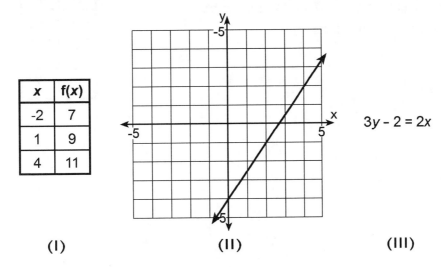

(I)   (II)   (III)

Part A

Which function has the greatest rate of change?

*Answer:*_____

Part B

Does any pair of functions have the same rate of change? *Justify your answer.*

Show your work:

*Answer:*_____

64 Write an equation of a linear function, in the form $f(x) = mx + b$, that represents the given table of values.

x	y
0	5
1	6
2	7

65 Solve the given problem by using a system of linear equations.

A man bought 3 shirts and 8 pairs of slacks for $237. At the same store, another man bought 5 shirts and 3 pairs of slacks for $147. Find the cost of 1 shirt and 1 pair of slacks.

TEST 4

1 What is the value of x for which the equation $6^{-7} \cdot 6^{-x} = \dfrac{1}{216}$ is true?

    A   4

    B   −3

    C   3

    D   −4   1 _____

2 $12^4 \cdot 12^5 \cdot 12 =$

    A 12^{10}

    B 12^9

    C 144^{20}

    D 12^{20} 2 _____

3 What is the value of $\sqrt{225}$?

    A   50,625

    B   112.5

    C   25

    D   15   3 _____

4 What is 2,040,000,000 written in scientific notation?

    A 2.4×10^8

    B 2.4×10^6

    C 2.04×10^9

    D 2.04×10^7 4 _____

5 If two angles of a triangle each measure 70°, the triangle is described as

    A   isosceles

    B   equilateral

    C   right

    D   obtuse   5 _____

6 Convert the given expression into scientific notation: $\dfrac{\left(5 \times 10^9\right)\left(8 \times 10^{-2}\right)}{2 \times 10^3}$

    A 2×10^9

    B 2×10^5

    C 2×10^3

    D 2×10^6 6 _____

7 In $\triangle ABC$, m$\angle A$ = 41° and m$\angle B$ = 48°. What kind of triangle is $\triangle ABC$?

A   Isosceles

B   Obtuse

C   Acute

D   Right   7 _____

8 Mandy's class and Peter's class have the same ratio of boys to girls. Mandy's class has 18 boys and 12 girls. If Peter's class has 15 boys, then how many girls does it have?

A   7

B   9

C   10

D   6   8 _____

9 Solve the equation for the given variable.
$$9y - 2y + 8 = 4y + 38$$

A   −12

B   10

C   12

D   −10   9 _____

10 Solve the given system of equations by substitution.
$$2x + 3y = 4$$
$$3x + 2y = 11$$

A   (−4, 8)

B   (2, 0)

C   (5, −2)

D   (−3, 10)   10 _____

11 The display of a student's calculator shows **2E3**. This is equivalent to

A   8

B   20,000

C   2,000

D   6   11_____

12 Write "5 ounces for $1,600" as a rate in simplest form.

A $\dfrac{1}{\$320}$

B $\dfrac{\$320}{1}$

C $\dfrac{5}{\$1,600}$

D $\dfrac{\$1,600}{5}$ 12 _____

13 The accompanying scatter plot shows the relationship between math grades and hours spent studying.

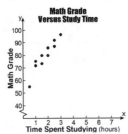

Which graph represents the line of best fit?

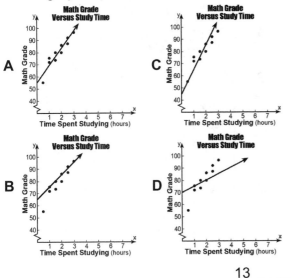

13 _____

14 Mrs. Jackson owns a large music store. She requires 2 supervisors for every 5 trainees. If she has 6 supervisors and no trainees, what is the maximum number of trainees she can hire?

A   30

B   12

C   6

D   15   14 _____

15 What type of transformation moves (x, y) to $(5x, 5y)$?

A   translation

B   rotation

C   reflection

D   dilation   15_____

16 What is an equation of a linear function that represents the following table of values?

x	y
2	4
3	5
4	6

A $g(x) = x - 2$

B $g(x) = x + 2$

C $g(x) = 2x$

D $g(x) = 2x - 1$ 16_____

17 What are the solutions to the equation $25x^2 = 4$?

A $x = \pm \frac{2}{5}$

B $x = \frac{5}{2}$, only

C $x = \frac{2}{5}$, only

D $x = \pm \frac{5}{2}$ 17 _____

18 The graphs of the equations in the system of equations $3x + y = 3$ and $y - x = 5$ will

A   intersect at exactly one point

B   intersect at all points on the line $y = -3x + 3$

C   intersect at exactly two points

D   not intersect   18 _____

19 The length of a rectangle is 5 feet more than the width. The perimeter is 22 feet. Which system of equations below will determine the length (ℓ) in feet and the width (w) in feet of the rectangle?

A $\ell = w + 5$ and $2\ell + 2w = 22$

B $w = \ell + 5$ and $2\ell + 2w = 22$

C $\ell - w = 5$ and $\ell + w = 22$

D $\ell = 5w$ and $22 = \ell + w$ 19 _____

20 Which graph illustrates the solution to $x + 2y = -4$ and $3x - 2y = 9$?

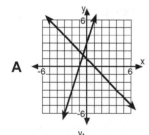
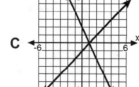
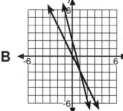
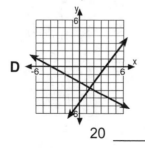

20 _____

21 Which of the following is not a representation of the function graphed below?

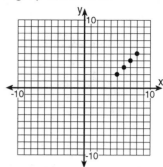

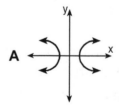
A

B $\{(x, y)| \; y = x - 3; \; x = 5, 6, 7, 8\}$

C

x	y
5	2
6	3
7	4
8	5

D   (2, 5), (3, 6), (4, 7), (5, 8)   21 _____

22 Which of the following graphs of a relation is also a function?

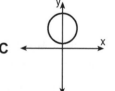
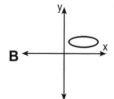
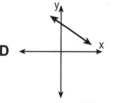

22 _____

23 Which of the following is the correct graphic representation of the linear function $g(x) = -x - 2$?

A 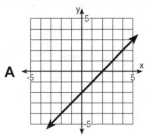 C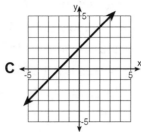

B 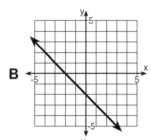 D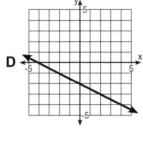

23 _____

24 If the measure of an angle is $12x$, which expression represents the measure of its supplement?

A $90 - 12x$

B $180 + 12x$

C $180 - 12x$

D $90 + 12x$

24 _____

25 Evaluate: $\left(\dfrac{2^{-2} - 2^{-1}}{2^{-1}}\right)^{-2}$

A $\dfrac{1}{4}$

B 4

C -4

D 15

25 _____

26 During a 45minute lunch period, Alonzo (A) went running and Brett (B) walked for exercise. Their times and distances are shown in the graph below.

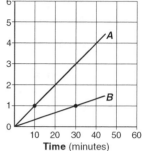

How much faster was Alonzo running than Brett was walking (in miles per hour)?

A 2 mph

B 4 mph

C 6 mph

D 10 mph

26 _____

27 Which of the following is the correct graphic representation of the function $g(x) = 2$?

A 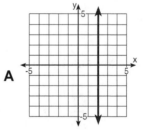 C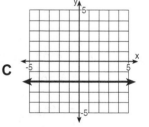

B 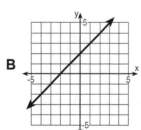 D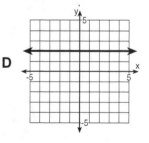

27 _____

28 What is an equation of the linear function that represents the following table of values?

x	y
1	7
2	7
3	7

A $x = 7$

B $y = 7$

C $y = x + 6$

D $y = x - 6$

28 _____

29 Which of the following is the best description of a dilation of a figure?

A   an enlargement or a reduction of the figure

B   a mirror image of the figure

C   a slide of the figure

D   a turning of the figure about some fixed point

29 _____

30 Which graph illustrates the solution to $x + y = 3$ and $x - y = 4$?

A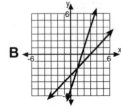

C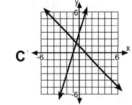

B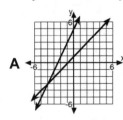

D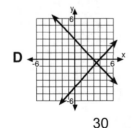

30 _____

31 In the story about the race between the tortoise and the hare, the tortoise moves at a very slow constantrate from start to finish. The hare, however, starts out very fast and is so far ahead that he decides to stop and take a nap. When he wakes up, the tortoise has passed him. He runs to catch up, but the tortoise crosses the finish line just ahead of the hare. The graph of which functions below correctly depicts this race between the tortoise and the hare?

KEY:
——— Tortoise
·········· Hare

A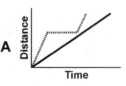

C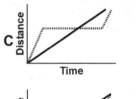

B

D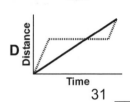

31 _____

32 If two angles of a triangle measure 43° and 48°, the triangle is

A   acute

B   isosceles

C   obtuse

D   right

32 _____

33 Under what type of transformation can the image be a different size than the original figure?

A   translation

B   dilation

C   rotation

D   reflection

33 _____

Test 4 – Part 1

34 The graph below indicates the speed of a toboggan (snow sled) as it travels away from the house.

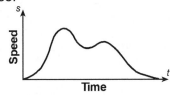

Which of the following statements best describes the relationship between speed and time?

A   There is not enough information given.

B   The toboggan (sled) is pulled up to the top of a steep hill, is ridden down the other side until it hits a small valley, goes over a small slope, and speeds down the other side until it coasts to a stop.

C   The toboggan (sled) starts out slowly and picks up speed as it moves down a hill, slows when going up a small slope, and picks up speed down the other side before coasting to a stop.

D   The toboggan (sled) moves farther away from the house, moves back toward the house, then farther away again, and approaches the starting point.   34 _____

35 If a straight line has a slope of zero, then for any two points on the line the change in y values is always equal to

A   0

B   −1

C   1

D   undefined   35 _____

36 A survey of high school students asked, "What is your favorite school lunch?". The results are shown in the two-way frequency table below.

'Favorite School Lunch' Poll

Grade	Lunch Choice				
	PIZZA	TACOS	CHICKEN	OTHER	TOTAL
9th	53	21	13	23	110
10th	41	11	9	44	105
11th	61	23	16	15	115
12th	38	29	9	19	95
TOTAL	193	84	47	101	425

According to the data in the table, what is the relative frequency of a 12th grade student's favorite lunch being either pizza or tacos?

A   0.73

B   0.158

C   0.40

D   0.705   36 _____

37 Which of the following properties of an object are preserved under a dilation?

A   size and shape, only

B   orientation and size, only

C   shape and orientation, only

D   size, shape, and orientation   37 _____

38 In the diagram, what type of transformation makes $\triangle A'B'C'$, the image of $\triangle ABC$?

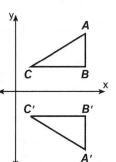

A   dilation

B   translation

C   rotation centered at the origin

D   reflection in the x-axis

                                38 _____

39 Line *m* is parallel to the *x*-axis, but not on the *x*-axis. In what two quadrants might the image of line *m* be located after reflecting over the *x*-axis?

A   II and III

B   I and III

C   III and IV

D   I and IV   39 _____

40 The measures of the acute angles of a right triangle are in the ratio 3:2. What is the measure of the smallest angle of the triangle?

A   144°

B   36°

C   54°

D   56°   40 _____

41 Which of the following is the correct graphic representation of the linear function $f(x) = -2x + 1$?

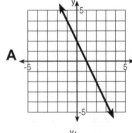
A

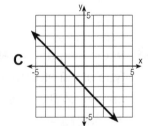
C

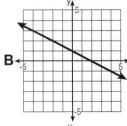
B

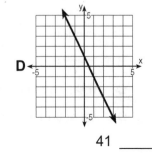
D

41 _____

42 What is the measure of the missing leg in the right triangle pictured below?

A   14

B   49

C   21

D   35

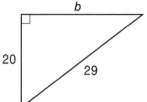

42 _____

43 Which of the following is the correct graphic representation of the linear function $h(x) = x + 5$?

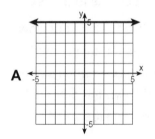
A

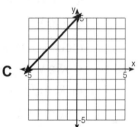
C

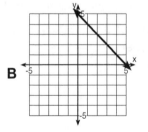
B

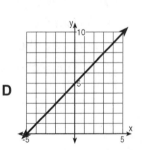
D

43 _____

Test 4 – Part 1

44 To get from his high school to his home, Jamal travels 5.0 miles east and 4.0 miles north. When Sheila goes to her home from the same high school, she travels 8.0 miles east and 2.0 miles south.

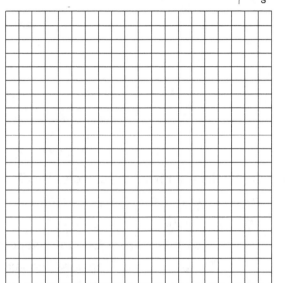

What is the measure of the shortest distance, to the nearest tenth of a mile, between Jamal's home and Sheila's home?

A 8.2 mi

B 6.4 mi

C 6.7 mi

D 7.0 mi   44 _____

45 Which of the following is a rational number and a terminating decimal?

A $2.\overline{7}$

B $\frac{1}{2}$

C $-\sqrt{7}$

D $\frac{1}{3}$ 45 _____

46 Chloe wants to decorate the entire outside surface of a cylindrical-shaped box with wallpaper to match her bedroom.

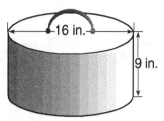

What is the minimum amount of wallpaper she will need to complete the job? Round your answer to the *nearest hundredth* of a square inch.

A 1,306.24 in.²

B 854.08 in.²

C 401.92 in.²

D 2,059.84 in.²   46 _____

47 Use the number line below to answer the given question.

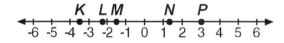

Which of the letters shown represents $-\frac{3}{2}$ on the number line?

A *M*

B *N*

C *K*

D *L*   47 _____

48 Which graph represents the strongest negative correlation?

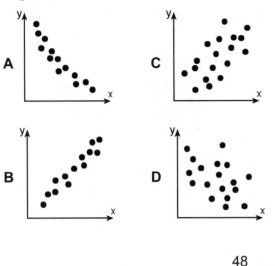

48 _____

49 Given the relation $R = \{(-2, 3), (d, 4), (1, 9), (0, 7)\}$, which replacement for d makes this relation a function?

A 4

B −2

C 1

D 0

49 _____

50 Express the given number or expression in scientific notation: 5,120,000

A 5.12×10^{6}

B 51.2×10^{5}

C 0.512×10^{6}

D 5.12×10^{-6}

50 _____

51 In the diagram below, rectangle *ABCD* was reflected over the *x*-axis to form rectangle *A'B'C'D'*. Then, rectangle *A'B'C'D'* was reflected over the *y*-axis to form rectangle *A"B"C"D"*. What single transformation could take rectangle *ABCD* to rectangle *A"B"C"D"*?

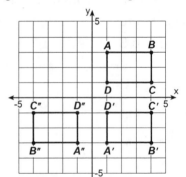

A D_1

B R_{180}

C $T_{(-5, -4)}$

D $r_{y = -x}$

51 _____

52 The two-way frequency table below shows the number of hours math students spent on homework and whether they worked by themselves or had help completing the assignment.

Math Homework

	Worked One Hour or Less	Worked More than One Hour
Worked Alone	45	25
Worked with Help	20	35

What is the relative frequency of math students who did their homework by themselves in reference to the total number of students in the survey?

A   0.52

B   0.56

C   0.357

D   0.44   52 _____

53 On Wednesday, Andrew got up late so he had to run to get to school on time. After school, he walked home and stayed there several hours until he returned to school for a band concert. After the concert, he walked straight home. The graph of which function below correctly depicts this situation?

A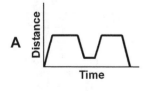

B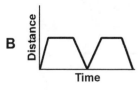

C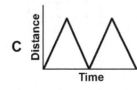

D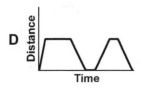

53 _____

Questions 54 and 55 refer to the following:

The accompanying table shows the relative frequency, based on the total number of people surveyed, of individuals who were asked "Do you own a tablet computer?".

Do You Own a Tablet Computer?

	Owns a Tablet	Does Not Own a Tablet	TOTAL
Males		0.16	
Females	0.34		0.62
TOTAL			

54 In reference to the total number of participants in the survey, what is the relative frequency that a male does not own a tablet computer?

A   0.44

B   0.38

C   0.16

D   0.22   54 _____

55 In reference to the total number of participants in the survey, what is the relative frequency of a female being surveyed?

A   0.62

B   0.28

C   0.44

D   0.56   55 _____

56 **Part A**

Given the function $-x - 4 = y$, complete the accompanying table by finding the missing input values, x, and the missing output values, y.

Part B

From the results obtained in Part A, plot a point on the grid for each of the x and y coordinates in the table. Connect the points with a line and label the line $-x - 4 = y$.

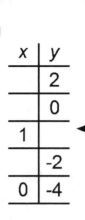

x	y
	2
	0
1	
	-2
0	-4

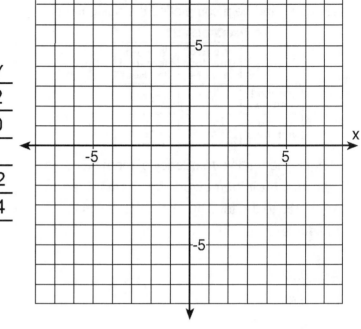

57 Triangle $A'B'C'$ is the image of triangle ABC after a dilation.

Determine the coordinates of the center of dilation and the scale of dilation.

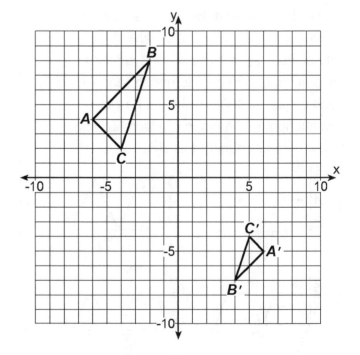

Questions 58 and 59 refer to the following:

Determine the solution to the system of equations whose graphs are shown:

58

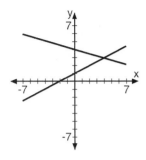

58 _____

59

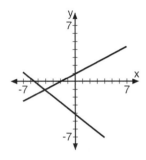

58 _____

60 Complete the function table below and graph the function. Determine if it is linear or nonlinear.
If it is a linear function, determine an equation for the function.

x	f(x)	(x,f(x))
-3	6	
-1	0	
0	0	
1	2	
3	12	

61 Use the formula V = *ℓwh* to calculate the volume of a rectangular solid.

Find the volume of the solid below.

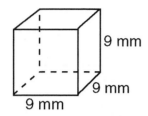

9 mm

9 mm

9 mm

62 Solve for *x*:   2(0.1x + 0.16) = 8.1

63 The three different linear functions shown below are represented by three different methods.

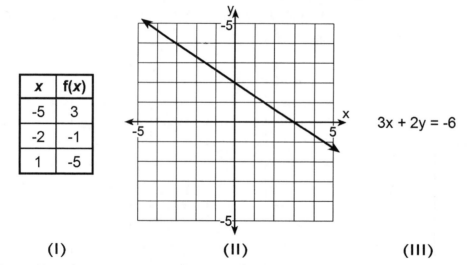

x	f(x)
-5	3
-2	-1
1	-5

3x + 2y = -6

Part A   (I)   (II)   (III)

Which function has a rate of change with the smallest numerical value?

Part B

Does any pair of functions have the same rate of change? *Justify your answer.*

64 A prepaid phone card charges a flat rate of 30 cents per minute.

Part A
Determine the unit rate of change for this proportional relationship.

Part B
In the accompanying table, list four additional ordered pairs that satisfy this proportional relationship.

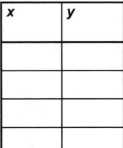

Part C
On the accompanying axes, plot these ordered pairs and sketch the line that represents this proportional relationship.

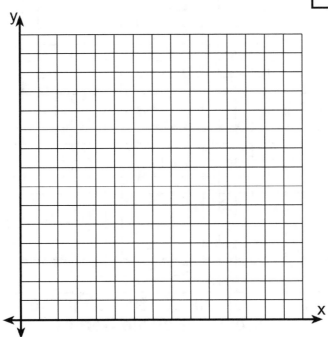

Part D
Determine the slope of the line graphed in Part C.

65 The amount of pancake batter needed, *B(x)*, for a community breakfast is a linear function of the number of people attending the breakfast, *x*. The table below shows the number of cups of batter required to serve the number of people indicated.

Number of People Served (x)	Cups of Batter Needed B(x)
10	7
20	14

Part A

Write an equation of this function, in the form *B(x) = mx + b*, and draw the graph of this function on the axes provided.

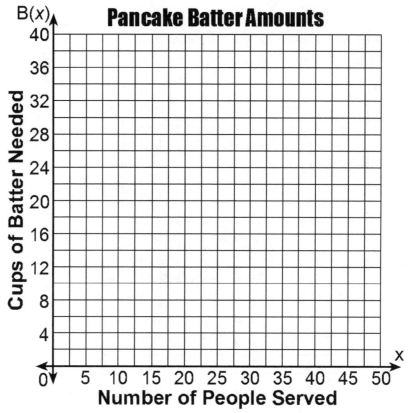

Part B

How many cups of batter must be prepared to serve 50 people?

*Answer:*_____

1 Which of the following numbers is irrational?

A $\sqrt{9}$

B π

C 16

D 2.1

1 _____

2 Which graph illustrates the solution to $3x - y = -5$ and $x - y = 1$?

A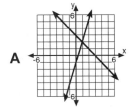

C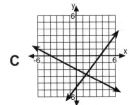

B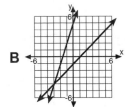

D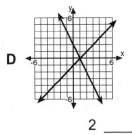

2 _____

3 Between which two consecutive whole numbers can $\sqrt{11}$ be found?

A 3 and 4

B 2 and 3

C 4 and 5

D 1 and 2

3 _____

4 $\sqrt{46}$ is between which two whole numbers?

A 45 and 47

B 7 and 8

C 6 and 7

D 6.7 and 6.8

4 _____

5 In the accompanying diagram, what type of transformation makes triangle 2 the image of triangle 1?

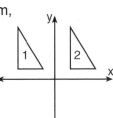

A dilation

B reflection in the *y*-axis

C rotation centered at the origin

D translation

5 _____

6 Solve the equation for the given variable.
$$10x + 8 = -32$$

A 4

B −40

C 24

D −4

6 _____

7 What is the value of $\sqrt{0.16}$?

   A 0.4

   B 0.2

   C 0.8

   D 0.04   7 _____

8 The average distance Mercury is from the Sun is approximately 57,900,000 km. The average distance Neptune is from the Sun is 4.497×10^9 km. Approximately how many times greater is Neptune's distance than Mercury's distance?

   A 780

   B 8

   C 78

   D 7,800   8 _____

9 Use the graph of the function to determine the corresponding positive input value for an output value of 4.

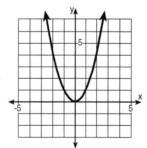

   A 1

   B 3

   C −2

   D 2   9 _____

10 If the number of molecules in 1 mole of a substance is 6.02×10^{23}, then the number of molecules in 100 moles is

   A 6.02×10^{21}

   B 6.02×10^{25}

   C 6.02×10^{22}

   D 6.02×10^{24} 10 _____

11 Write "$9.16 for 4.3 pounds" as a unit rate. Round to the nearest cent.

   A $2.14/pound

   B $2.41/pound

   C $2.31/pound

   D $2.13/pound   11_____

12 The accompanying circle graph shows how the Marino family spends its income each month.

What is the measure, in degrees, of the central angle that represents the percentage of income spent on food?

   A 90°

   B 25°

   C 50°

   D 360°   12 _____

13 $3^2 \cdot 3^7 \cdot 3^0 =$

   A 3^0

   B 3^9

   C 27^0

   D 27^9 13 _____

14 For the accompanying equation, which step would not be a possible first step for solving this equation algebraically?

$$\frac{5}{8}(3x - 2) - 2\frac{1}{4} = 6 - \frac{1}{3}x$$

   A   adding $2\frac{1}{4}$ and 6

   B   multiplying -2 by $\frac{5}{8}$

   C   adding $\frac{1}{3}x$ and $3x$

   D   multiplying every term of the equation by 24   14 _____

15 Which of the following expressions is equivalent to $\frac{9}{16}$?

   A $4^{-2} \times 3^2$

   B $4^2 \times 3^{-2}$

   C $4^{-2} \times 3^{-2}$

   D $\left(\frac{3}{4}\right)^{-2}$ 15 _____

16 Which of the following numbers has the greatest value?

$2.78 \times 10^7, \quad 2.85 \times 10^9, \quad 2.96 \times 10^8, \quad 2.85 \times 10^8$

   A 2.78×10^7

   B 2.85×10^9

   C 2.96×10^8

   D 2.85×10^8 16 _____

17 Which graph illustrates the solution to $2x - y = -5$ and $x - y = -1$?

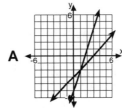 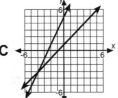

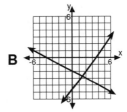 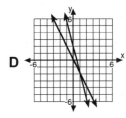

                                   17 _____

18 Solve the given system of equations by substitution. $y = 2x$
                              $x + y = 12$

   A $(15, -3)$

   B $(2, 4)$

   C $(6, 6)$

   D $(4, 8)$ 18 _____

19 The following system of equations intersect at one point. $y = 12x - 1$
$$3x - 2y = 1$$

What is the x coordinate of the point of intersection of the linear equations?

A   1

B   −1

C $-\dfrac{1}{2}$

D $\dfrac{3}{2}$ 19_____

20 Convert the given expression into scientific notation: $\dfrac{(5 \times 10^9)(8 \times 10^2)}{2 \times 10^{-3}}$

A 2×10^{15}

B 2×10^9

C 2×10^{14}

D 2×10^8 20 _____

21 Deshi invested $32,000. He invested part of it at 8% interest per year and the rest at 6% interest per year. If his total interest income for the year was $2,120, then how much was invested at 6%?

A $22,000

B $16,000

C $12,000

D $10,000   21_____

22 Use the graph of the function below to determine the corresponding output value for an input value of 2.

A   −2

B   −3

C   7

D   5

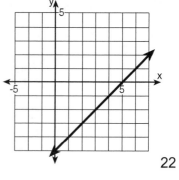

22 _____

23 Given the function f(x) = {(3, 1),(6, 2), (9, 3), (12, 4)}, which of the following is another way to represent this function?

A $y = \dfrac{1}{3}x$

B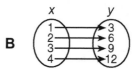

C $y = 3x$ and the domain is (1, 2, 3, 4}

D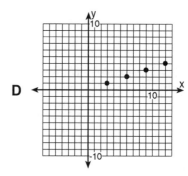

23 _____

24 What is an equation of the linear function that represents the following table of values?

x	y
1	-1
2	-2
3	-3

A $g(x) = 1 - x$

B $g(x) = x$

C $g(x) = -x$

D $g(x) = x - 1$ 24 _____

25 The cross-section of a drainage ditch is an isosceles triangle. The ditch is 4 feet wide at ground level and 3 feet deep.

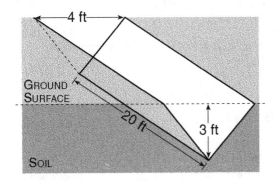

How many cubic feet of soil was removed to dig a ditch 20 feet long?

A   90 ft³

B   120 ft³

C   240 ft³

D   180 ft³   25 _____

26 What is an equation of the linear function whose graph passes through the points (−10, 6) and (15, −14)?

A $f(x) = \frac{4}{5}x - 2$

B $f(x) = -2x - \frac{4}{5}$

C $f(x) = -\frac{5}{4}x - 2$

D $f(x) = -\frac{4}{5}x - 2$ 26 _____

27 A translation maps $A(1, 2)$ onto $A'(-1, 3)$. What are the coordinates of the image of the origin under the same translation?

A   (−1, 2)

B   (0, 0)

C   (−2, 1)

D   (2, −1)   27 _____

28 What is an equation of the linear function whose graph passes through the points (−1, −12) and (5, 6)?

A $f(x) = 9x - 3$

B $f(x) = \frac{1}{3}x - 9$

C $f(x) = 3x - 9$

D $f(x) = -9x + 3$ 28 _____

29 The accompanying table shows the relative frequency, based on the total number of people surveyed, of individuals who were asked "Do you own a tablet computer?".

Do You Own a Tablet Computer?

	Owns a Tablet	Does Not Own a Tablet	TOTAL
Males		0.16	
Females	0.34		0.62
TOTAL			

In reference to the total number of participants in the survey, what is the relative frequency that an individual does not own a tablet computer?

A 0.44

B 0.16

C 0.28

D 0.38

29 _____

30 What is an equation of the linear function that represents the following table of values?

A $h(x) = x + 2$

B $h(x) = x - 2$

C $h(x) = x - 1$

D $h(x) = -x$

x	y
1	-1
2	0
3	1

30 _____

31 The cost, c, of parking a car in a lot in Lowville is given by the formula $c = 0.75h + 1.50$, where h is the number of hours parked. If Sara paid $3.75 for parking in a lot downtown, how long was she parked?

A 2.25 hours

B 3 hours

C 5 hours

D 1.5 hours

31 _____

32 What is an equation of the linear function that represents the following table of values?

A $y = x + 4$

B $y = x - 4$

C $x = y + 4$

D $y = 4x$

x	y
1	5
2	6
3	7

32 _____

33 The graph demonstrates the relationship between the distance a person bicycles and the time elapsed.

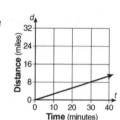

What is the speed of this bicyclist?

A 16 mph

B 32 mph

C 8 mph

D 24 mph

33 _____

34 What is an equation of the linear function that represents the following table of values?

A $h(x) = x - 2$

B $h(x) = x + 3$

C $h(x) = 2x + 1$

D $h(x) = x + 2$

x	y
1	3
2	5
3	7

34 _____

Use the graph below to answer questions 35 and 36.

The rate at which water is entering water tank for any time $t > 0$ is shown in the graph below. A positive rate indicates that water is flowing into the tank, and a negative rate indicates that water is leaking from the tank.

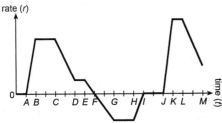

35 What is the longest interval of time for which the volume of water in the tank is decreasing?

    A   *EH*

    B   *EG*

    C   *FG*

    D   *FH*   35 _____

36 Over which interval of time is the rate of water flow changing the fastest?

    A   *LM*

    B   *AB*

    C   *KL*

    D   *JK*   36 _____

37 What transformation does not always produce an image that is congruent to the original figure?

    A   translation

    B   rotation

    C   dilation

    D   reflection   37 _____

38 Since dilations preserve shape and not size, which of the following is the correct conclusion?

    A   dilations preserve similarity

    B   dilations preserve congruence

    C   dilations preserve similarity and congruence

    D   dilations do not preserve similarity   38 _____

39 What type of transformation is illustrated in the accompanying diagram?

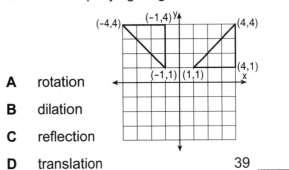

    A   rotation

    B   dilation

    C   reflection

    D   translation   39 _____

40 Which one of the following diagrams represents a reflection in line ℓ?

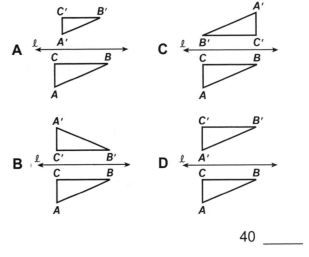

        40 _____

41 In the accompanying diagram, which point is the image of point *K* after a line reflection in the *x*-axis?

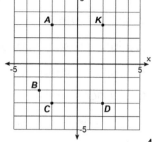

A *A*

B *B*

C *C*

D *D*

41 _____

42 Reflect triangle *QRS* over the *y*-axis and label the image appropriately.

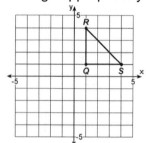

What are the coordinates of point *R'*?

A (−4, 1)

B (−1, −4)

C (1, 4)

D (−1, 4)

42 _____

43 If Willy is traveling due north on his skateboard and he does a 90° counterclockwise rotation, in which direction is he now traveling?

A north

B south

C west

D east

43 _____

44 If trapezoid *RSTV* below was reflected over the line *y* = −1, which of the following graphs would represent the image of trapezoid *RSTV*?

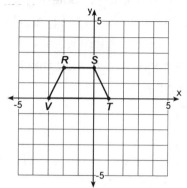

A 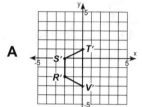   **C**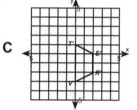

B 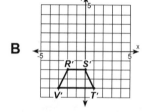   **D**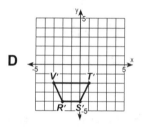

44 _____

45 A dilation with a scale factor of $\frac{1}{5}$ would be equivalent to a dilation with a scale factor of _____.

A 0.1

B 0.2

C 0.4

D 0.5

45 _____

Test 5 – Part 1

46 Which of the following graphs is the pre-image of square F'G'H'J' before a dilation with a scale factor = 2.5 centered at the origin?

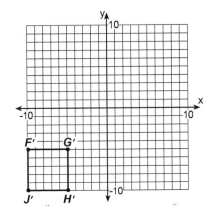

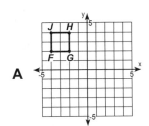

A

C

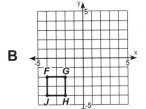
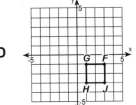

B

D

46 _____

47 If a figure completely contained in quadrant I is dilated with a scale factor of −2, in what quadrant will the image be located?

A   III

B   II

C   I

D   IV

47 _____

48 If the measures of the three angles of a triangle are represented by x°, (2x − 20)°, and (3x − 10)°, then the triangle is

A   obtuse

B   right

C   equilateral

D   isosceles

48 _____

49 In the accompanying diagram, parallel lines $\overleftrightarrow{AB}$ and $\overleftrightarrow{CD}$ are cut by transversal $\overleftrightarrow{FE}$ at points G and H, respectively.

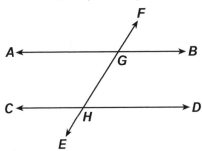

If m∠BGH = (3x)° and m∠CHF = (7x − 164)°, what is m∠BGH?

A   133°

B   33°

C   57°

D   123°

49 _____

50 In $\triangle ABC$, m$\angle A = 41°$ and m$\angle B = 48°$. What kind of triangle is $\triangle ABC$?

A right

B obtuse

C acute

D isosceles

50 _____

51 Angle A is five times the measure of angle B. Angles A and B are interior angles on the same side of the transversal. Angles A and C are alternate interior angles. What is the measure of angle C?

A 60°

B 150°

C 145°

D 30°

51 _____

52 In the figure below, $\overline{AB} \cong \overline{BC}$.

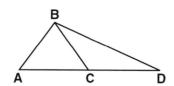

If m$\angle BCD = 126°$, what is m$\angle ABC$?

A 126°

B 54°

C 108°

D 72°

52 _____

53 Which of the following triangles shows the measures of a right triangle?

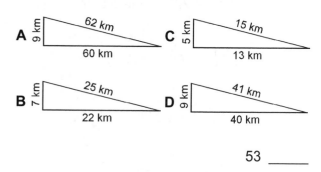

53 _____

54 In the diagram below, two squares are placed together to begin forming a right triangle.

If a third square is added as shown, what will be the length of a side of this square?

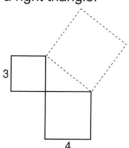

A 25

B 4

C 5

D 6

54 _____

55 What is the length of the line segment whose endpoints are (1, 1) and (3, −3)?

A $\sqrt{20}$

B $\sqrt{32}$

C 10

D $\sqrt{8}$

55 _____

Test 5 – Part 1

56 The length of a rectangular picture frame is 5 inches less than 3 times the width. If the perimeter of the frame is 78 inches, find the dimensions of this frame.

Show your work:

*Answer:*_____

57 A sequence composed of a dilation and a congruency transformation maps $\triangle ABC$ onto its image $\triangle A'B'C$.

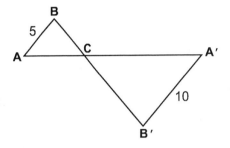

Part A

Identify the type of congruency transfomation.

Part B

Determine the scale factor k.

58 On the axes provided, sketch a graph of the given equation. Determine an appropriate scale for each axis and label the graph with its equation.

$y = 4x$

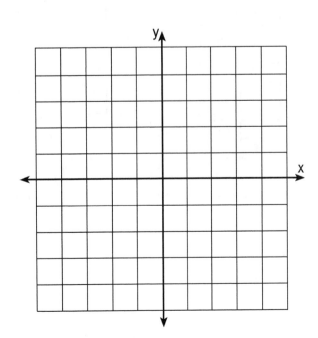

59 What is the volume of the sphere shown, expressed in terms of π, if radius (*r*) equals 6 inches?

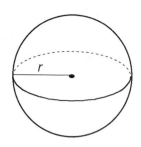

60 Use the accompany graph to do the following:

Part A

Construct a scatter plot for the data in the table below.

x	y
0	8
0.5	9
1.0	7
2.0	5
0.5	8
0	9
2.0	4
2.5	2
1.5	7
3.0	-1
1.5	6
1.0	8
2.5	1
3.0	-2

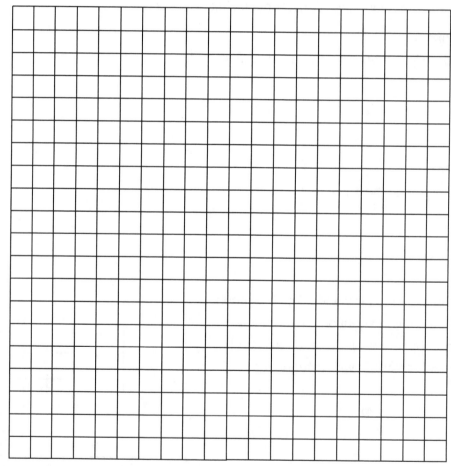

Part B

Describe the correlation that exists between the variables *x* and *y*.

61 Two different tests were designed to measure understanding of a topic. The two tests were given to ten students with the following results.

Test x	75	78	88	92	95	67	58	72	74	81
Test y	81	73	85	88	89	73	66	75	70	78

Part A

Construct a scatter plot for these scores, and, using your calculator, find an equation for the line of best fit. Round slope and intercept to the *nearest hundredth*.

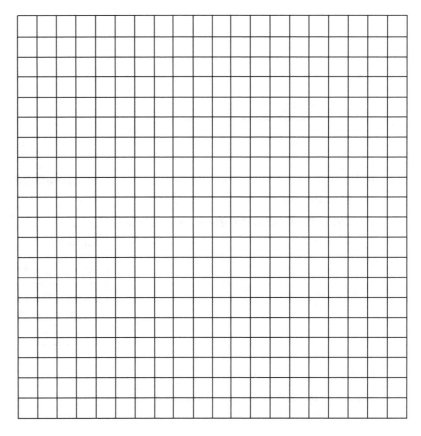

Part B

Interpret the *y*-intercept of the line of best fit in the context of this situation.

Part C

Predict the score, to the nearest integer, on test *y* for a student who scored 87 on test *x*.

Show your work:

Answer: _____

62 The first of two lines passes through the points (−3, 7) and (9, 4). The second line passes through the points (−8, −3) and (8, −7). Determine algebraically whether these two lines intersect, are parallel, or are coinciding lines. If the lines intersect, state the coordinates of the point they have in common.

Show your work:

*Answer:*_____

63 In the accompanying table, partial data for the 30 games played by the middle school soccer team is recorded.

Middle School Soccer Game Statistics

	Home Games	Away Games	TOTAL
Games Won		6	
Games Lost	6	5	
Games Tied	4		
TOTAL		14	30

Part A

How many away games ended in a tie? _____

Part B

What is the percentage of home games won? _____

Part C

What is the percentage of games lost? _____

Part D

How many games were played at home? _____

Part E

How many games did not end in a tie? _____

64 Customers at Bargain World were asked whether they liked the new pricing policy the store had recently instituted. The data is recorded in the accompanying frequency table.

Customers' Opinions on New Prices

	Like	Dislike	TOTAL
Males	60	120	180
Females	240	80	320
TOTAL	300	200	500

Part A

Construct a two-way relative frequency that displays the relative frequencies in reference to gender of the survey participants.

Part B

What is the relative frequency of a male liking the new prices?

*Answer:*_____

Part C

What is the relative frequency of a female disliking the new prices?

*Answer:*_____

65 A factory is producing and stockpiling metal sheets to be shipped to an automobile manufacturing plant. The factory ships only when there is a minimum of 2,050 sheets in stock. The accompanying table shows the day, x, and the number of sheets in stock, $f(x)$.

Day (x)	Sheets in Stock $(f(x))$
1	860
2	930
3	1,000
4	1,150
5	1,200
6	1,360

Part A

Using a calculator, find an equation for the line of best fit for this set of data, rounding the coefficients to four decimal places.

Part B

Use the equation written in Part A to determine the day the sheets will be shipped. *Show all work.*

Show your work:

*Answer:*_____

Part C

Interpret the the meaning of the slope in this situation.

Test 5 – Part 2

TEST 6

1 $\frac{5}{8}$ is a(n)

    A   irrational number

    B   integer

    C   natural number

    D   rational number   1 _____

2 Which graph illustrates the solution to $2x + y = 2$ and $x - y = 1$?

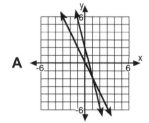
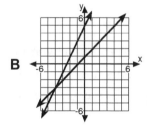

             2 _____

3 Between which two consecutive whole numbers can $\sqrt{29}$ be found?

    A   4 and 5

    B   6 and 7

    C   5 and 6

    D   3 and 4   3 _____

4 Which integer is closest to $\sqrt{18}$?

    A   18

    B   4.2

    C   5

    D   4   4 _____

5 In the accompanying diagram, the faces are congruent. What type of transformation is illustrated?

    A   a reflection in line ℓ

    B   a translation

    C   a rotation

    D   a dilation   5 _____

6 Solve the equation for the given variable.
$$5x + 22 = -73$$

    A   −15

    B   −19

    C   −12

    D   −10   6 _____

7 Find the length of the radius of a sphere whose volume is 16.

A $\sqrt[3]{\dfrac{12}{\pi}}$

B $\sqrt[3]{12\pi}$

C $\sqrt[3]{\dfrac{16}{\pi}}$

D $\sqrt[3]{16\pi}$

7 _____

8 What number has a value that is one one-thousandth the value of 3.7×10^{-2}?

A 3.7×10^{-5}

B 0.0037×0.01^{-2}

C 3.7×10^{-6}

D 0.0037×10^{-2}

8 _____

9 Use the graph of the function to determine the corresponding positive input value for an output value of –2.

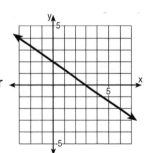

A 3

B –3

C 6

D 4

9 _____

10 The letters on the graph below represent four vehicles traveling at different rates from the same location. In the distance formula $d = rt$, r represents the rate of change, or slope.

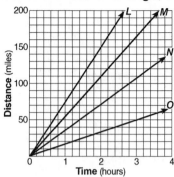

Which ray on the given graph best represents a slope of 35 mph?

A N

B M

C O

D L

10 _____

11 7 out of 12 students surveyed enjoyed playing basketball. If 696 students are surveyed, how many would you expect to say they enjoy playing basketball?

A 399

B 406

C 198

D 290

11 _____

Test 6 – Part 1

12 $7^5 \cdot 7^0 \cdot 7 =$

A 7^0

B 27^9

C 7^6

D 27^0 12 _____

13 The interest on a loan varies directly as the rate. If the rate is halved, then the interest

A   is halved

B   is doubled

C   remains the same

D   is multiplied by 4   13 _____

14 For the accompanying equation, which step would not be a possible first step for solving this equation algebraically?

$$\frac{2}{3}(2x-1) + \frac{7}{3} = 7 + \frac{1}{2}x$$

A   multiplying every term of the equation by six

B   multiplying -1 by $\frac{2}{3}$

C   subtracting $\frac{7}{3}$ from 7

D   subtracting $\frac{1}{2}x$ from $2x$ 14 _____

15 Which of the following expressions is equivalent to $\frac{1}{25}$?

A $5^{-2} \times 5^4$

B $(0.5)^{-2}$

C $(5^{-4})^2$

D $5^{-1} \times 5^{-1}$ 15 _____

16 Which of the following numbers has the greatest value?

A 4.509×10^7

B 4.51×10^7

C 4.509×10^6

D 4.51×10^6 16 _____

17 Which graph illustrates the solution to $3x - y = 1$ and $x + y = 3$?

A 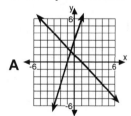   C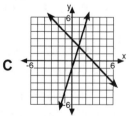

B 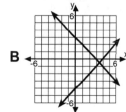   D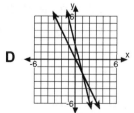

                              17 _____

18 Solve the given system of equations by substitution. $x - y = 3$
$$8x + 3y = 13$$

A   (5, 2)

B $(-\frac{1}{4}, 5)$

C   (2, –1)

D   (–2, –5)   18 _____

19 What is an equation of the linear function that represents the following table of values?

A $g(x) = x + 2$

B $g(x) = 2x + 2$

C $g(x) = x - 3$

D $g(x) = x + 3$ 19 _____

x	y
1	4
2	6
3	8

20 Jon and Sue both walk for exercise. The graph shows how far each of them walked during 50 minutes of exercise. In miles per hour, how much greater is Jon's rate of walking than Sue's rate?

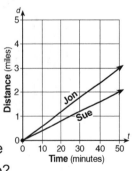

A   1 mph

B   1.5 mph

C   2 mph

D   1.2 mph   20 _____

21 Joan bought a total of 30 cans of cola and root beer. There were twice as many cans of cola as cans of root beer. Which system of equations will determine the number of cans of cola (c) and the number of cans of root beer (r) that Joan bought?

A $c + r = 30$ and $r = 2c$

B $c + r = 30$ and $c = 2r$

C $r = 30 + c$ and $c = 2r$

D $c = r + 30$ and $r = 2c$ 21 _____

22 Use the graph of the function to determine the corresponding output value for an input value of 4.

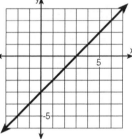

A   0

B   1

C   7

D   –1   22 _____

23 In the function $y = -3x + 5$, what is the output value, y, when the input value is $x = 1$?

A   2

B   5

C   –5

D   –2   23 _____

24 What is an equation of the linear function that represents the following table of values?

x	y
1	3
2	4
3	5

- **A** $h(x) = x + 2$
- **B** $h(x) = 2x$
- **C** $h(x) = -2x$
- **D** $h(x) = x - 2$

24 _____

25 Use the formula $V = 12bh\ell$ to calculate the volume of a triangular prism.

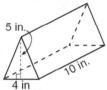

5 in.
10 in.
4 in

What is the volume of the solid?

- **A** 100 in.³
- **B** 50 in.³
- **C** 200 in.³
- **D** 400 in.³

25 _____

26 If a linear function is always increasing then its rate of change is

- **A** negative
- **B** zero
- **C** positive
- **D** undefined

26 _____

27 Billy starts at the point (2, −4). He travels due north 5 units, due west 3 units, and then due north again 4 units.

What are the coordinates of his final destination?

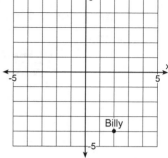

- **A** (−1, 5)
- **B** (−5, 1)
- **C** (5, −1)
- **D** (1, −5)

27 _____

28 If a linear function is a constant function then its rate of change is

- **A** positive
- **B** zero
- **C** undefined
- **D** negative

28 _____

29 The two-way frequency table below shows the number of hours math students spent on homework and whether they worked by themselves or had help completing the assignment.

Math Homework

	Worked One Hour or Less	Worked More than One Hour
Worked Alone	45	25
Worked with Help	20	35

What is the relative frequency of math students who worked I hour or less in reference to the total number of students in the survey?

A   0.168

B   0.56

C   0.52

D   0.36   29 _____

30 What is an equation of the linear function that represents the following table of values?

A $h(x) = x + 3$

B $h(x) = 4$

C $h(x) = x + 2$

D $h(x) = 4x$

x	y
0	4
3	4
6	4

30 _____

31 If a linear function is always decreasing then its rate of change is

A   zero

B   positive

C   undefined

D   negative   31 _____

32 What is an equation of the linear function that represents the following table of values?

A $y = x + 1$

B $y = x$

C $x = -x$

D $y = x - 1$

x	y
1	0
2	1
3	2

32 _____

33 Solve the given system of equations by substitution. $a = b - 2$
$$2a + 3b = 21$$

A   (−2, 0)

B   (0, 2)

C   (4, 6)

D   (3, 5)   33_____

34 What is an equation of the linear function that represents the following table of values?

A $f(x) = -2x$

B $f(x) = x - 6$

C $f(x) = x - 3$

D $f(x) = 2x$

x	y
1	-2
2	-4
3	-6

34 _____

35 Use the number line below to answer the given question.

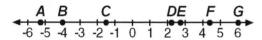

Which of the letters shown represents $\sqrt{5}$ on the number line?

A E

B F

C D

D C

35 _____

36 Between which two consecutive whole numbers can $\sqrt{113}$ be found?

A 9 and 10

B 10 and 11

C 11 and 12

D 56 and 57

36 _____

37 Under what type of transformation is size not preserved?

A rotation

B reflection

C translation

D dilation

37 _____

38 Triangle $A'B'C'$ is the image of $\triangle ABC$ under a given transformation. If $\triangle A'B'C'$ is similar but not congruent to $\triangle ABC$, the transformation must be a _____.

A dilation

B line reflection

C translation

D rotation

38 _____

39 Graph the 90 degree clockwise rotation of trapezoid *BDFA* about the origin and label the image appropriately.

Which point in the rotated image matches the location of point *F* in the original?

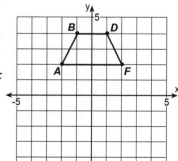

A D'

B B'

C A'

D F'

39 _____

40 A translation moves $P(3, 5)$ to $P'(6, 1)$. What are the coordinates of the image of point $(-3, -5)$ under the same translation?

A $(-5, -3)$

B $(-6, -9)$

C $(-6, -1)$

D $(0, -9)$ 40 _____

41 If segment $\overline{PQ}$ is reflected over the y-axis, in what quadrants do the image points lie?

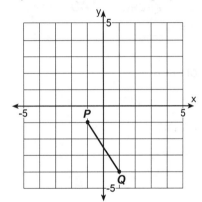

A   I and IV, only

B   III and IV, only

C   I and II, only

D   III, only   41 _____

42 If the hour and minute hand on an analog clock read 7:15, what would be the approximate time if each of the hands were rotated 180° clockwise?

A   12:30

B   1:30

C   1:45

D   12:45   42 _____

43 Graph the reflection of rectangle $RSTV$ over the line $y = x$ and label the image appropriately.

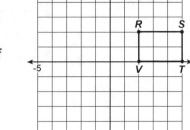

What are the coordinates of point R'?

A $(2, 2)$

B $(0, 2)$

C $(0, 5)$

D $(2, 5)$ 43 _____

44 If segment $\overline{YZ}$ below was reflected over the x-axis, which of the following graphs would represent the image of segment $\overline{YZ}$?

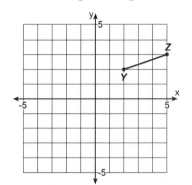

A 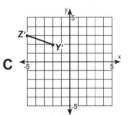   C

B 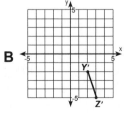   D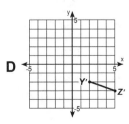

44 _____

Test 6 – Part 2

45 What is the center of dilation for the image of △*ABC* in the given diagram?

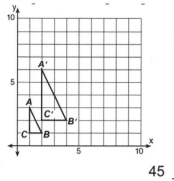

A   (1, 1)

B   (1, 2)

C   (0, 0)

D   (2, 1)

45 _____

46 Which of the following graphs is the pre-image of triangle *A'B'C'* before a dilation with a scale factor = 3 centered at the origin?

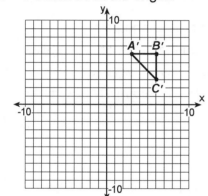

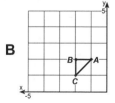

A

B

C

D

46 _____

47 If a figure completely contained in quadrant I is dilated with a scale factor of >0, in which quadrant will the image be located?

A   IV

B   II

C   III

D   I

47 _____

48 Given: In △*PQR*, $\overline{PQ} \cong \overline{QR}$.
If m∠*Q* = 50°, what is m∠*P*?

A   130°

B   65°

C   50°

D   25°

48 _____

49 When two parallel lines are intersected by a transversal, four interior angles are formed. The measures of two alternate interior angles are given as 3(*x* + 20)° and (23*x*)°. What are the measures of the two angles?

A   111° and 69°

B   69° and 21°

C   69° and 69°

D   23° and 63°

49 _____

50 In $\triangle QRS$, $m\angle Q = x°$, $m\angle R = (8x - 40)°$, and $m\angle S = 2x°$. Which type of triangle is $\triangle QRS$?

A   obtuse

B   acute

C   right

D   isosceles   50 _____

51 In the accompanying diagram, parallel lines $\overleftrightarrow{XY}$ and $\overleftrightarrow{MN}$ are cut by transversal $\overleftrightarrow{TS}$ at points A and B, respectively.

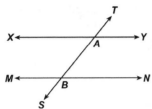

If $m\angle XAS = (5x)°$ and $m\angle NBT = (9x - 40)°$, what is $m\angle NBT$?

A   10°

B   59°

C   139°

D   50°   51 _____

52 In the figure below, $\overline{AB} \cong \overline{BC}$.

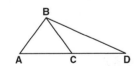

If $m\angle BCD = 128°$, what is $m\angle ABC$?

A   104°

B   128°

C   76°

D   52°   52 _____

53 Which of the following triangles shows the measures of a right triangle?

53 _____

54 In the diagram below, three squares are placed to form the sides of $\triangle ABC$.

If the area of square R is 36 and the area of square S is 64, what is the area of square T?

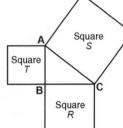

A   81

B   28

C   121

D   100   54 _____

55 A circle has its center at (4, 2) and passes through the point (9, 6). What is the approximate length of the diameter? Round your answer to the *nearest tenth*.

A   6.0

B   12.8

C   9.4

D   6.4   55 _____

56 The perimeter of a rectangle is 86 cm. The length of this rectangle is 2 less than 4 times the width. Find, in centimeters, the length and width of the rectangle.

Show your work:

*Answer:*_____

57 A sequence composed of a dilation and a congruency transformation maps △ABC onto its image △A'B'C'.

Part A

Identify the type of congruency transformation.

*Answer:*_____

Part B

Determine the scale factor *k*.

*Answer:*_____

58 On the axes provided, sketch a graph of the given equation. Determine an appropriate scale for each axis and label the graph with its equation.

Equation: $y = -2x$

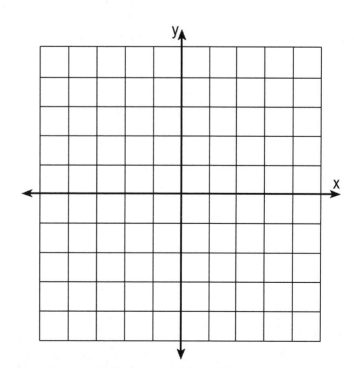

59 A farmer feeds his animals twice a day from a rectangular trough filled to the top. The trough is 3 ft 4 in. long, 2 ft wide, and 6 in. high.

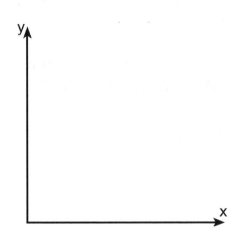

Assuming the animals completely empty the trough of food, how many cubic inches of food do his animals eat each day?

A 5,760 in.^{3} **B** 10,368 in.^{3} **C** 2,880 in.^{3} **D** 11,520 in.^{3} 59_____

60 Draw an example of a scatter plot that shows a strong negative linear correlation with one outlier.

61 The relationship of a woman's shoe size and length of a woman's foot, in inches, is given in the table below.

Shoe Size	5.0	6.0	7.0	8.0
Foot Length (in.)	9.0	9.25	9.5	9.75

Construct a scatter plot and determine an equation for the line of best fit.

62 Find the area of triangular region in the first quadrant bounded by the *y*-axis and the lines $2x - y = -2$ and $3x + 2y = 18$. *Use of the accompanying grid is optional.*

Show your work:

*Answer:*_____

63 A high school surveyed 1,200 students in grades 9 through 12 asking them to identify their favorite subject in school. A two-way relative frequency table based on the total number of survey participants was created, as shown to the right.

Students' Favorite School Subjects

	Math	Science	Social Studies	English
Freshmen	0.05	0.08	0.12	0.11
Sophomores	0.04	0.10	0.06	0.08
Juniors	0.06	0.09	0.03	0.02
Seniors	0.03	0.06	0.02	0.05

Part A

How many students like English? _____

Part B

How many freshmen like Math? _____

Part C

How many more sophomores like English than Math? _____

Part D

How many students like English and Social Studies combined? _____

Part E

How many juniors were surveyed? _____

Part F

How many freshmen were surveyed? _____

Part G

Create a two-way frequency table with the given data.

Part A

Given the function $\frac{1}{2}x - y = 8$, complete the values

for x and y in the accompanying table.

x	y
	-2
	4
5	
0	
	-4

Part B

From the results obtained in Part A, plot a point
on the grid for each of the x and y coordinates
in the table. Connect the points with a line and
label the line $\frac{1}{2}x - y = 8$.

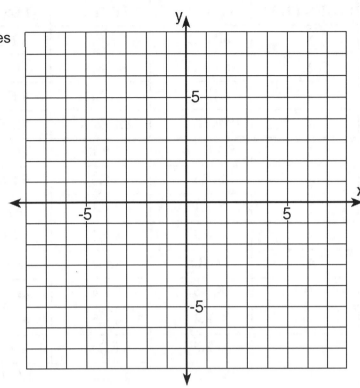

65 Fill in the accompanying table with the perimeter and area of a square using the given various lengths
for a side.

Side Length of a Square	1	3	4	5	7	8	10	12	15
Perimeter of a Square									
Area of a Square									

Does either the perimeter or area row of the table represent a linear function? If yes, write an equation for
this linear function.

Correlation of Standards

QUESTION	TEST 1	TEST 2	TEST 3	TEST 4	TEST 5	TEST 5
1	8.EE.2	8.EE.2	8.EE.1	8.EE.1	8.NS.1	8.NS.1
2	8.EE.3	8.EE.3	8.EE.1a	8.EE.1a	8.EE.8b	8.EE.8b
3	8.EE.4	8.EE.4	8.EE.2	8.EE.2	8.NS.2	8.NS.2
4	8.EE.7a	8.EE.7a	8.EE.3	8.EE.3	8.NS.2	8.NS.2
5	8.EE.8b	8.EE.8b	8.G.A.2	8.G.A.2	8.G.3	8.G.3
6	8.F.3	8.F.3	8.EE.4	8.EE.4	8.EE.7a	8.EE.7a
7	8.F.1	8.F.1	8.GA.2	8.GA.2	8.EE.2	8.EE.2
8	8.F.3	8.F.3	8.EE.5	8.EE.5	8.EE.3	8.EE.3
9	7.GB.5	8.F.5	8.EE.7b	8.EE.7b	8.F.1	8.F.1
10	8.F.5	8.F.4	8.EE.8a	8.EE.8a	8.EE.4	8.EE.4
11	8.F.4	8.F.5	8.EE.4	8.EE.4	8.EE.5	8.EE.5
12	8.F.5	8.EE.8c	8.SP.2	8.EE.4 & 5	8.EE.5	8.EE.1
13	8.EE.8c	8.EE.1	8.EE.4 & 5	8.SP.2	8.EE.1	8.EE.5
14	8.EE.1	8.EE.4	8.EE.4	8.EE.4	8.EE.7b	8.EE.7b
15	8.EE.4	7.GB.5	8.G.4	8.G.4	8.EE.1	8.EE.1
16	7.GB.5	8.F.4	8.F.4	8.F.4	8.EE.3	8.EE.3
17	8.F.4	8.EE.1	8.F.4	8.EE.7	8.EE.1b	8.EE.1b
18	8.EE.1	8.EE.3	8.EE.8b	8.EE.8b	8.EE.8b	8.EE.8b
19	8.EE.3	8.EE.4	8.EE.8c	8.EE.8c	8.EE.8b	8.F.4
20	8.EE.4	8.SP.3	8.EE.8a	8.EE.8a	8.EE.4	8.EE.4
21	8.SP.3	8.SP.4	8.F.1a	8.F.1a	8.EE.8b	8.EE.8b
22	8.EE.4	8.SP.2	8.F.1	8.F.1	8.F.1	8.F.1
23	8.EE.1	8.SP.3	8.F.2	8.F.2	8.F.1	8.F.1
24	8.SP.2	7.GA.2	7.G.5b	7.G.5b	8.F.4	8.F.4
25	8.SP.3	8.G.5	8.F.2	8.EE.1	8.G.9	8.G.9
26	7.GA.2	8.EE.1	8.EE.1	8.F.2	8.F.4	8.F.4
27	8.G.5	8.G.2	8.F.3	8.F.3	8.G.3	8.G.3
28	8.G.2	7.GB.5	8.F.4	8.F.4	8.F.4	8.F.4
29	8.EE.5	8.EE.5	8.G.1a	8.G.1a	8.SP.4	8.SP.4
30	8.G.9	8.G.9	8F.5b	8.EE.8b	8.F.4	8.F.4
31	8.G.3	8.G.4	8.F.5	8F.5b	8.EE.7b	8.F.4
32	8.G.4	8.G.3	8.NS.1	7.GA2	8.F.4	8.F.4
33	8.G.1.a	8.G.1.a	8.EE.8b	8.G.4	8.EE.5	8.EE.8b

Correlation of Standards

QUESTION	TEST 1	TEST 2	TEST 3	TEST 4	TEST 5	TEST 6
34	8.G7	8.G7	7.G.A.2	8.G.5	8.F.4	8.F.4
35	8.EE.6	8.EE.6	8.EE.6	8.EE.6	8.F.5	8.F.5
36	8.EE.8b	8.EE.8b	8.SP.4	8.SP.4	8.F.5	8.F.5
37	8.G.8	8.G.8	8.G.1a	8.G.1a	8.G.1a	8.G.1a
38	8.EE.6	8.EE.6	8.NS.2	8.G.3	8.G.1a	8.G.1a
39	8.EE.7a	8.EE.7a	8.G.3	8.G.3	8.G.3	8.G.3
40	8.EE.7b	8.EE.7b	8.G.3	8.G.5	8.G.3	8.G.3
41	8.EE.8b	8.EE.8b	8.G.4	8.F.3	8.G.3	8.G.3
42	7.GA.3	8.G.1.a	8.G.5	8.G.7	8.G.3	8.G.3
43	8.G.1.a	7.GA.3	8.F.3	8.F.3	8.G.3	8.G.3
44	8.G.2	8.G.2	8.G.7	8.G.8	8.G.3	8.G.3
45	8.F.5	8.F.5	8.F.3	8.NS.1	8.G.4	8.G.4
46	8.G.9	8.G.9	8.G.8	8.G.9	8.G.4	8.G.4
47	8.EE.6	8.G.2	8.G.9	8.NS.2	8.G.4	8.G.4
48	8.G.2	8.EE.5	8.SP.1	8.SP.1	8.G.5	8.G.5
49	8.EE.5	8.EE.8a	8.F.A.1	8.F.A.1	8.G.5	8.G.5
50	8.G.9	8.EE.6	8.SP.4	8.EE.4	8.G.5	8.G.5
51	8.EE.8a	8.EE.7b	8.G.9	8.G.9	8.G.5	8.G.5
52	8.SP.4	8.SP.4	8.SP.4	8.SP.4	8.G.5	8.G.5
53	8.EE.7b	8.F.5	8.F.5	8.F.5	8.G.7	8.G.7
54	8.F.5	8.G.9	8.SP.4	8.SP.4	8.G.7	8.G.7
55	8.G.3	8.G.3	8.SP.4	8.SP.4	8.G.8	8.G.8
56	8.EE.7b	8.EE.7b	8.F.1	8.F.1	8.EE.7b	8.EE.7b
57	8.F.3	8.F.3	8.G.4	8.G.4	8.G.4	8.G.4
58	8.EE.8b	8.EE.8b	8.EE.C.8a	8.EE.C.8a	8.EE.5	8.EE.5
59	8.F.2	8.F.2	8.F.3	8.F.3	8.G.9	8.G.9
60	8.EE.7b	8.EE.7b	8.G.9	8.G.9	8.EE.7b	8.EE.7b
61	8.F.1	8.F.1	8.EE.7b	8.EE.7b	8.SP.3	8.SP.3
62	8.EE.8c	8.EE.8c	8.EE.5	8.F.2	8.EE.8c	8.EE.8c
63	8.G.3	8.G.3	8.F.2	8.EE.5	8.SP.4	8.SP.4
64	8.G.4	8.G.4	8.F.4	8.F.4	8.SP.4	8.SP.4
65	8.EE.5	8.EE.5	8.EE.8c	8.EE.8c	8.SP.3	8.SP.3